A Natural History of Rape

A Natural History of Rape
Biological Bases of Sexual Coercion

Randy Thornhill
Craig T. Palmer

The MIT Press
Cambridge, Massachusetts
London, England

Second printing, 2000

Set in Sabon.
Printed and bound in the United States of America.

Library of Congress Cataloging-in-Publication Data

Thornhill, Randy.
A natural history of rape: biological bases of sexual coercion / Randy Thornhill, Craig T. Palmer
p. cm.
Includes bibliographical references and index.
ISBN 0-262-20125-9 (alk. paper)
1. Rape. 2. Men—Sexual behavior. 3. Human evolution.
I. Palmer, Craig T. II. Title
HV6558.T48 2000
364.15'32—dc21 99-31685
CIP

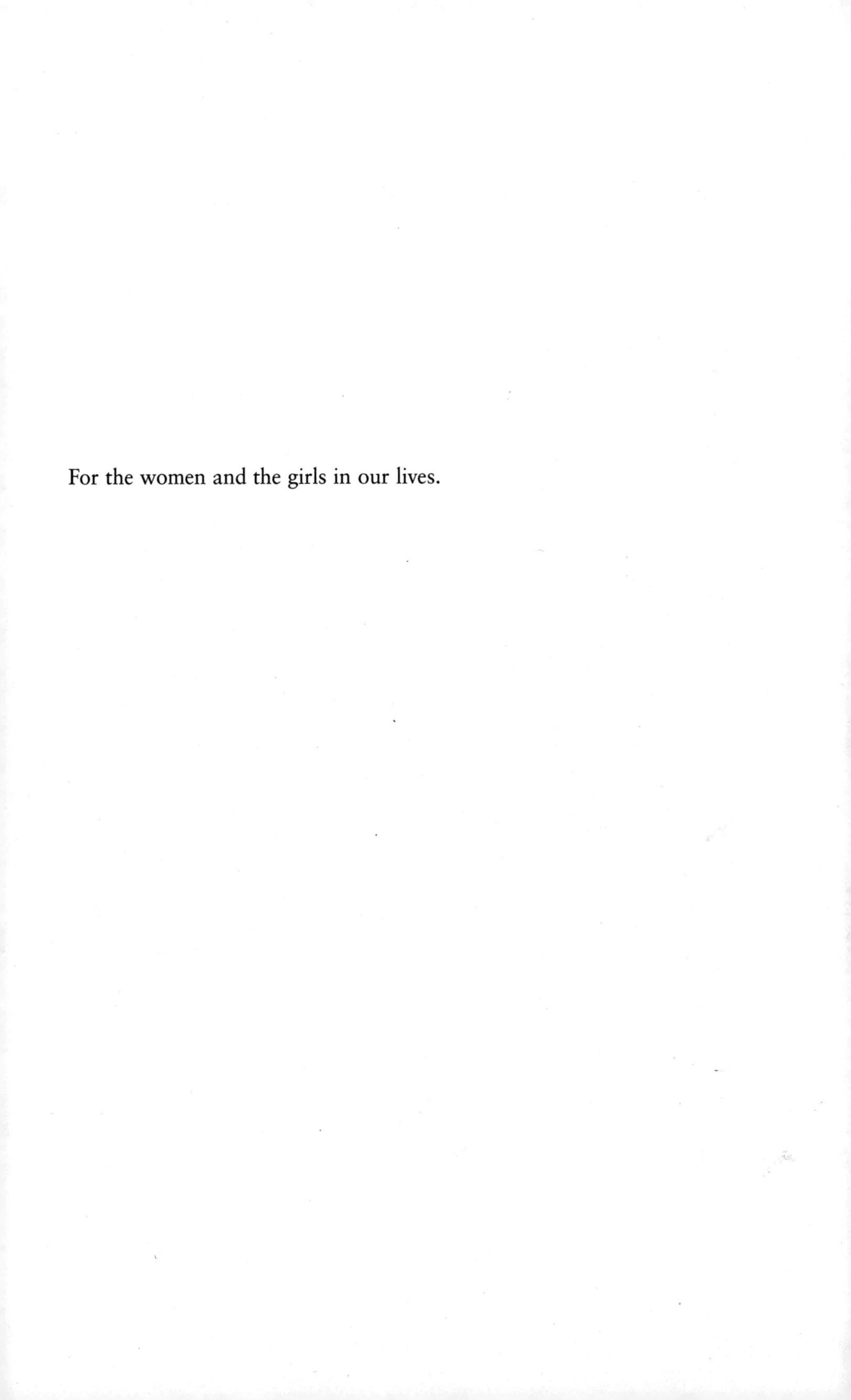

For the women and the girls in our lives.

Contents

Foreword

Rape is horrific for women. The mere thought of rape arouses anxiety, revulsion, and anger, so it is not surprising that women are very ambivalent about subjecting rape to scientific scrutiny. Research on deadly or grotesquely disfiguring diseases probably arouses less distaste and ambivalence. Ambivalence arises out of the mix of anxiety, revulsion, and a desire to have more information.

The authors of *A Natural History of Rape* are familiar with various expressions of such ambivalence, and they understand why women are so anxious. As scientists, they value knowledge and assume that trying to understand everything about why rape occurs is far more beneficial for women in the long term even if the scientific inquiry inspires anxiety and revulsion. The evolution-minded scientific approach that the authors espouse has resulted in many novel and nonintuitive insights about why rape occurs and why women are so devastated by the victimization. Thornhill and Palmer are doubly handicapped by the topic and by their theoretical approach since most people have no relevant background in evolutionary biology. Even evolutionary biologists can be reluctant to think about our own species as if it were just another (very special) species.

The evolution-minded approach that Thornhill and Palmer have developed with respect to rape is carefully articulated in this book. The programmatic overview applies to other aspects of human affairs, including psychopathology. Their assumptions, the logic of their inferences, and their standards of evidence are explicitly laid out for the interested reader. They do not engage in subterfuge or bafflegab. Any weaknesses in the data or assumptions or inferences are there to be discerned, and therein lies their intellectual strength. They passionately embrace the scientific method whereby claims about causal inferences are open to public scrutiny, revision, and better understanding. Neither science, nor Darwinism, nor any other -ism has a unitary voice. The authors themselves are not always in agreement, but their alternate hypotheses and inferences can be resolved with further scientific investigation. The authors are strongest when discussing why rape occurs and why it is so harmful to women. The intellectual confrontations with those who do not appreciate the power of scientific methods or their conceptual framework have obviously inspired the authors to listen and think hard about the misunderstandings, but neither author has undertaken a systematic investigation of the history and sociology of interdisciplinary communication and competition. Nevertheless, their insights may serve others well.

Thornhill and Palmer are among a growing number of evolutionists whose research and scholarship are producing new knowledge about the causes and consequences of various kinds of interpersonal harms, and what might prove to be effective prevention policies.

Margo Wilson

Preface

As scientists who would like to see rape eradicated from human life, we hold that the ability to accomplish such a change is directly correlated with how much is known about the causes of human behavior. In contrast, mistaken notions about what causes rape are almost guaranteed to hinder its prevention.

Unfortunately, there is, as the zoologist Patricia Gowaty (1997, p. 1) puts it, a "troublesome antipathy of modern society, including many feminists, to science and scientific discourse." "This scientific illiteracy," Gowaty continues, "has led to shallow understandings of the nature of science and ignorance of basic Darwinian processes." Others have made similar points about the low value placed on science and about the profound misunderstanding of science in some academic circles.[1] Indeed, it has been suggested that "most of the influential work in the social sciences is ideological" (Leslie 1990, p. 896). Whether or not this is true of the social sciences as a whole, there is evidence that many of the rape-prevention strategies proposed during the past three decades relied upon explanations of rape based more on ideology than on scientific evidence.

Reviewing social scientists' writings on human sexual coercion published in the period 1982–1992, the psychologists Del Thiessen and Robert Young (1994, p. 60) conclude that "the messages seem more political than scientific." Similarly, we find that the majority of the researchers on whose theories today's attempts to solve the problem of rape are based remain uninformed about the most powerful scientific theory concerning living things: the theory of evolution by Darwinian selection. As a result, many of the social scientists' proposals for dealing with rape are based on assumptions about human behavior that have been without theoretical justification since 1859, when Charles Darwin's book *On the Origin of Species* was published.

Some say that the rape researchers' ignorance of evolutionary principles is unimportant because the solutions to rape offered by evolutionarily informed researchers are, as the biologist and women's studies professor Zuleyma Tang-Martinez (1996, p. 122) puts it, "all solutions that could have been arrived at by a feminist psychosocial analysis, without invoking evolutionary biology." This claim is incorrect. Not only does an evolutionary approach generate new knowledge that could be used to decrease the incidence of rape; some of the proposals put forth by individuals uninformed by evolutionary theory may actually *increase* it.

We realize that our approach and our frankness will trouble some social scientists, including some serious and well-intentioned rape researchers. We also suspect that more of those researchers might consider the evolutionary approach if its compatibility with the social sciences' current explanations were to be emphasized more, and we applaud the recent attempts by other evolutionists to reach out to them; indeed, we predict that those efforts will eventually establish the evolutionary approach as the paradigm for the social sciences. In the meantime, however, an approach that minimizes or overlooks inaccuracies in the social sciences' present explanations of rape has the cost of letting those inaccurate assumptions remain as bases for practical decisions. Police officers, lawyers, teachers, parents, counselors, convicted rapists, potential rapists, and children are being taught "rape-prevention" measures that will fail because they are based on fundamentally inaccurate notions about human nature. We hope that our criticism of what we call "the social science explanation" of rape will be understood in this light.

As Carl Hempel (1959), Karl Popper (1968), and other philosophers have argued, science progresses through the falsification of erroneous ideas. Scientific critiques, then, must focus on the very heart of the perceived difficulty with an idea or body of research. To show that a tangential or trivial part of some work is wrong and then argue that the work is fundamentally flawed is not valid scientific criticism. Similarly, to point out the errors of some investigators in a scientific field is not to demonstrate that the field is fundamentally flawed. Basing an attack on a trivial aspect of a work or on the weak research program of a field—a rhetorical approach, common in the humanities—is not valid in science.[2]

The social science theory of rape is based on empirically erroneous, even mythological, ideas about human development, behavior, and psychology. It contradicts fundamental knowledge about evolution. It fails to yield a coherent, consistent, progressive body of knowledge. The literature it has produced is largely political rather than scientific.

As the biologist James Lloyd (1979, p. 18) points out, Darwinism "provides a guide and prevents certain kinds of errors, raises suspicions of certain explanations or observations, suggests lines of research to be followed, and provides a sound criterion for recognizing significant observations on natural phenomena." In contrast, as the evolutionary anthropologist Donald Symons (1987a, p. 135) notes, research in the social sciences lacks the steadying guide of tested theory: "Because they have developed almost entirely innocent of Darwinism, the social and behavioral sciences have committed certain kinds of errors, put forth certain suspect explanations, failed to pursue certain lines of research, and by and large, lacked a sound criterion for recognizing significant observations."

Acknowledgments

This book is a true collaboration. Both of us worked on every page. The sequence of our names was determined randomly.

Many people helped us, but special recognition must be given to our research assistant, Scott Wright, who, in addition to performing a significant part of the literature research for the project, tirelessly read and made critical comments on multiple drafts of the manuscript. The book would not have been completed without his contribution. For additional critical comments on the entire manuscript we are very grateful to John Alcock, Paul Andrews, Paul Bethge, Laura Betzig, Amy Brand, David DiBari, Owen Jones, Donald Symons, Robert Trivers, and the four anonymous readers.

Michael Fisher, Marc Hauser, and Kirk Jensen provided useful comments on sections of the manuscript. Anne Rice expertly assisted with manuscript preparation and Joy Thornhill with indexing. R.T. thanks the Harry Frank Guggenheim Foundation for financial support of his studies of the psychological pain of rape victims.

Others have written persuasively of the power of the evolutionary approach to human rape. Although they met with only modest success, each attempt to show that biology has something important and fundamental to say about rape made it easier for the next. In 1979, Richard Alexander, Katherine Noonan, Richard Hagen, and Donald Symons made the first major attempts to place human rape in a modern evolutionary perspective. In 1983 William and Lea Shields published their important paper on the evolution of human rape. Del Thiessen wrote the first book-length manuscript on human rape and evolution in the mid 1980s, but he gave up on publishing it after failing to get it signed by any of the numerous publishers he had contacted. Fortunately, the atmosphere changed, and in 1989 Lee Ellis, who had written a number of scientific papers on the biology of rape, was able to get a book on human rape published. In the last decade, Vernon Quinsey and Martin Lalumiére managed to introduce an evolutionary perspective to studies of sex offenders. Over the same period, Jack Beckstrom and Owen Jones introduced the modern evolutionary perspective on rape into the scholarly literature of law. We celebrate the efforts of all these pioneers.

Everyone (except those who cannot distinguish a diagnosis of adaptation from a moral exhortation) could read it with interest.

—Mark Ridley, on jacket of Donald Symons's book *The Evolution of Human Sexuality* (1979)

1

Rape and Evolutionary Theory

Not enough people understand what rape is, and, until they do . . . , not enough will be done to stop it.
—rape victim, quoted in Groth 1979 (p. 87)

By one intuitive and relevant definition, rape is copulation resisted to the best of the victim's ability unless such resistance would probably result in death or serious injury to the victim or in death or injury to individuals the victim commonly protects.[1] Other sexual assaults, including oral and anal penetration of a man or a woman under the same conditions, also may be called rape under some circumstances.

In one study, 13 percent of the surveyed American women of ages 18 and older reported having been the victim of at least one completed rape—rape having been defined as "an event that occurred without the woman's consent, involved the use of force or threat of force, and involved sexual penetration[2] of the victim's vagina, mouth or rectum" (Kilpatrick et al. 1992, p. i). Other surveys using slightly different definitions or different data-collection procedures have found high rates too, especially when the survey procedures have given researchers access to victims of alleged rapes not reported to the police. Kilpatrick et al. (ibid., p. 6) estimate the percentage of rapes of women not reported at between 66 and 84. Of women who had experienced a rape involving penile-vaginal intercourse, from 37 to 57 percent experienced post-traumatic stress syndrome afterward—a frequency higher than that associated with any other crime against women, including aggravated assault, burglary, and robbery (Kilpatrick et al. 1987; Resnick et al. 1993).

We suggest two answers to the question of why humans have not been able to put an end to rape:

- Most people don't know much about why humans have the desires, emotions, and values that they have, including those that cause rape. This is because most people lack any understanding of the ultimate (that is, evolutionary) causes of why humans are the way they are. This lack of understanding has severely limited people's knowledge of the exact proximate (immediate) causes of rape, thus limiting the ability of concerned people to change the behavior.
- For 25 years, attempts to prevent rape have not only failed to be informed by an evolutionary approach; they have been based on explanations designed to make ideological statements rather than to be consistent with scientific knowledge of human behavior.

One cannot understand evolutionary explanations of rape, much less evaluate them, without a solid grasp of evolutionary theory. Failure to appreciate this point has caused much valuable time to be wasted on misplaced attacks on evolutionary explanations.

Assuming that the main interest of most readers of this book is the subject of rape rather than evolutionary theory per se, we now present some questions about rape that an evolutionary approach can answer:

- Why are males the rapists and females (usually) the victims?
- Why is rape a horrendous experience for the victim?
- Why does the mental trauma of rape vary with the victim's age and marital status?
- Why does the mental trauma of rape vary with the types of sex acts?
- Why does the mental trauma of rape vary with the degree of visible physical injuries to the victim, but in a direction one might not expect?
- Why do young males rape more often than older males?
- Why are young women more often the victims of rape than older women or girls (i.e., pre-pubertal females)?
- Why is rape more frequent in some situations, such as war, than in others?
- Why does rape occur in all known cultures?
- Why are some instances of rape punished in all known cultures?
- Why are people (especially husbands) often suspicious of an individual's claim to have been raped?
- Why is rape often treated as a crime against the victim's husband?

- Why have attempts to reform rape laws met with only limited success?
- Why does rape exist in many, but not all, species?
- Why does rape still occur among humans?
- How can rape be prevented?

Evolutionary Theory

The question "What is man?" is probably the most profound that can be asked by man. It has always been central to any system of philosophy or of theology. We know that it has been asked by the most learned humans 2000 years ago, and it is just possible that it was being asked by the most brilliant australopithecines 2 million years ago. The point I want to make now is that all attempts to answer that question before 1859 are worthless and that we will be better off if we ignore them completely. —Simpson 1966, p. 472

Intelligent life on a planet comes of age when it first works out the reason for its own existence. If superior creatures from space ever visit Earth, the first question they will ask, in order to assess the level of our civilization, is: "Have they discovered evolution yet?" Living organisms had existed on Earth, without ever knowing why, for more than three billion years before the truth finally dawned on one of them. His name was Charles Darwin. To be fair, others had inklings of the truth, but it was Darwin who first put together a coherent and tenable account of why we exist. —Dawkins 1976, p. 1

Many social scientists (and others) have dismissed claims such as these as evidence of a somehow non-scientific "messianic conviction" (Kacelnik 1997, p. 65). Although these quotes indicate considerable enthusiasm, the important question is whether they accurately describe the implications of the theory of evolution by natural selection. Simpson's and Dawkins's enthusiasm is warranted by the tremendous success of evolutionary theory in guiding the scientific study of life in general and of humans in particular to fruitful ends of deep knowledge.

Cause, Proximate and Ultimate

A friend of ours once told us that after a movie she returned with her date to his car in an isolated parking lot. Then, instead of taking her home, the man locked the doors and physically forced her to have sexual intercourse with him. The question addressed in this book, and the question asked us by our friend, is: What was the cause of this man's behavior?

In both the vernacular sense and the scientific sense, *cause* is defined as that without which an effect or a phenomenon would not exist. Biologists study two levels of causation: *proximate* and *ultimate*. Proximate causes of behavior are those that operate over the short term—the immediate causes of behavior. These are the types of causes with which most people, including most social scientists, are exclusively concerned. For example, if, when reading our friend's question concerning the cause of the man's behavior, you said to yourself it was because he hated women, felt the need to dominate someone, had been abused as a child, had drunk too much, had too much testosterone circulating in his body, was compensating for feelings of inadequacy, had been raised in a patriarchal culture, had watched too much violence on television, was addicted to violent pornography, was sexually aroused, hated his mother, hated his father, and/or had a rare violence-inducing gene, you proposed a proximate cause of his behavior. You probably didn't ask why your proposed proximate cause existed in the first place. That is, you probably didn't concern yourself with the ultimate cause of the behavior.

Because they refer to the immediate events that produce a behavior or some other phenotypic (i.e., bodily) trait, proximate causes include genes, hormones, physiological structures (including brain mechanisms), and environmental stimuli (including environmental experiences that affect learning). Proximate explanations have to do with *how* such developmental or physiological mechanisms cause something to happen; ultimate explanations have to do with *why* particular proximate mechanisms exist.

Proximate and ultimate explanations are complements, not alternatives. For example, the claim that millions of years of selection caused the human eye to have its current form (an ultimate explanation) is in no way contradictory to the claim that a series of rods and cones enable the eye to relay visual information to the brain (a proximate explanation). Similarly, the claim that learning affects men's rape behavior (i.e., that it is a proximate cause) does not contradict the view that the behavior has evolved.

Identifying ultimate causes, however, is important, because certain proximate explanations may be incompatible with certain ultimate explanations. This is because certain ultimate explanations specify the existence of certain types of proximate mechanisms. For example, the ultimate explanation that the human eye evolved by natural selection because it increased our ancestors' ability to detect light requires the existence of proximate light-detection mechanisms in the eye.

No aspect of life can be completely understood until both its proximate and its ultimate causation are fully known. To understand how ultimate causes can be known, one must understand how natural selection leads to *adaptations*.

Natural Selection and Adaptations

Adaptations are phenotypic features (morphological structures, physiological mechanisms, and behaviors) that are present in individual organisms because they were favored by natural selection in the past. Darwin sought to explain the existence of adaptation in terms of evolution by selection. Initially, he observed the action of selection on living things in nature—a fact of natural history that is inescapable in view of the high rates of reproduction and mortality in all organisms. Later, he realized just how creative selection could be when extended over the long history of life on Earth. This retrospection is evident in the following eloquent passage from *On the Origin of Species*:

> Natural selection is daily and hourly scrutinizing, throughout the world, every variation, even the slightest; rejecting that which is bad, preserving and adding up all that is good; silently and insensibly working. . . . We see nothing of these slow changes in progress, until the hand of time has marked the long lapse of ages. (Ridley 1987, p. 87)

The biologist George Williams, in his 1966 book *Adaptation and Natural Selection*, clarified what Darwin meant when he wrote of natural selection's rejecting all that was "bad" and preserving all that was "good." First, Williams noted, these words were not used in a moral sense; they referred only to the effects of traits on an individual's ability to survive and reproduce.[3] That is, "good" traits are those that promote an individual's reproductive interests. We evolutionists use the term *reproductive success* to refer to these reproductive interests, by which we mean not the mere production of offspring but the production of offspring that survive to produce offspring (Palmer and Steadman 1997). A trait that increases this ability is "good" in terms of natural selection even though one might consider it undesirable in moral terms. There is no connection here between what is biological or naturally selected and what is morally right or wrong. To assume a connection is to commit what is called the *naturalistic fallacy*. In addition, Williams clarified that natural selection favors traits that are "good" in the sense of increasing an individual's reproductive success, not necessarily traits that are "good" in the sense of increasing a group's ability to survive.

The idea that selection favors traits that increase group survival, known as *group selection*, had become very popular before the publication of Williams's book—especially after the publication of *Animal Dispersion in Relation to Social Behavior*, an influential book by the ornithologist V. C. Wynne-Edwards (1962). Williams's rebuttal of the concept of group selection convinced almost every biologist who read it that Wynne-Edwards was mistaken. However, the idea that selection favors traits that function for the good of the group appears to have been too attractive for many non-scientists to give up. Not only does it remain popular among the general public; it continues to have a small following among evolutionary biologists (Wilson and Sober 1994; Sober and Wilson 1998).[4]

One cannot grasp the power of natural selection to "design" adaptations until one abandons both the notion that natural selection favors traits that are morally good and the notion that it favors traits that function for the good of the group. Only then can one appreciate the power of natural selection to design complex traits of individuals.

The human eye's many physiological structures exist because they increased the reproductive success of individuals in tens of thousands of past generations. Although there are four agents of evolution (that is, four natural processes that are known to cause changes in gene frequencies of populations), selection is the only evolutionary agent that can create adaptations like the human eye. The other evolutionary agents (mutation, drift, and gene flow[5])—cannot produce adaptations; they lack the necessary creativity, because their action is always random with regard to environmental challenges (e.g., predators) facing individuals. Selection, when it acts in a directional, cumulative manner over long periods of time, creates complex phenotypic designs out of the simple, random genetic variation generated by the three other evolutionary agents. Selection is not a random process; it is differential reproduction of individuals by consequence of their differences in phenotypic design for environmental challenges. An adaptation, then, is a phenotypic solution to a past environmental problem that persistently affected individuals for long periods of evolutionary time and thereby caused cumulative, directional selection. Evolution by selection is not a purposive process; however, it produces, by means of gradual and persistent effects, traits that serve certain functions—that is, adaptations.

Adaptations do not necessarily increase reproductive success in current environments if those environments differ significantly from past environments. The seeds of a tree that fall on a city sidewalk are complexly designed adaptations, formed by selection over many generations in past environments, yet they have essentially no chance of surviving or reproducing in the current environment of the sidewalk. Similarly, the North American pronghorn antelope shows certain social behaviors and certain locomotory adaptations (e.g., short bursts of high speed) for avoiding species of large cats and hyenas that are now extinct (Byers 1997).

The difference between current and evolutionary historical environments is especially important to keep in mind when one is considering human behavioral adaptations. Today most humans live in environments that have evolutionarily novel components. (Modern contraception is one such component that obviously influences the reproductive success of individuals in an evolutionarily novel way.) Therefore, human behavior is sometimes poorly adapted (in the evolutionary sense of the word) to current conditions.

Evolutionary functional explanations also differ from the non-evolutionary functional explanations familiar to most social scientists. In fact, evolutionary functional explanations overcome a problem that has plagued non-evolutionary functional explanations. Non-evolutionary functional explanations are unable to explain why a particular trait has come to serve a certain function when alternative traits could also serve that function (Hempel 1959). For example, Emile Durkheim, one of the founders of sociology, tried to explain religion by stating that it functioned to maintain the social group (Durkheim 1912). That explanation, however, is unable to account for why religion, instead of numerous alternative institutions (e.g., political governments, non-religious social organizations and ideologies), fulfills this particular function. The concept of evolution by natural selection helps overcome this problem. *Any* gene that *happens* to arise by random mutation, and *happens* to have the effect of increasing an organism's reproductive success, will become more frequent in future generations. Eventually, additional random mutations will also *happen* to occur in future generations and will also be favored by natural selection. Over time, this process results in functionally designed traits. Randomness (in the form of mutations) and the non-random process of natural selection combine to answer the question of why a particular trait has evolved instead of other imaginable traits that conceivably could have served the same function.

There is also the important fact that selection works only in relation to what has already evolved. The process does not start anew each time. Thus, there are many features that seem poorly designed relative to what might be imagined as a better solution. For example, the crossing of the respiratory and digestive tracts in the human throat can cause death from choking on food. It would be better design—much safer in terms of survival—if our air and food passages were completely separate. However, all vertebrates (backboned animals) from fishes to mammals on the phylogenetic tree (the tree connecting all life to a common ancestor) have crossing respiratory and digestive tracts. The human respiratory system evolved from portions of the digestive system of a remote invertebrate ancestral species, and the food and air passages have been linked in non-functional tandem ever since (Williams 1992). The crossing of passages is a historical legacy of selection's having built respiratory adaptations from ancestral digestive system features. Not itself an adaptation, it is a by-product of selection's having molded respiratory adaptation from what came before.

Similarly, any new mutation, through its bodily effect, is assessed by selection in relation to how well it performs in the evolved environment of other individuals in the population as well as in the evolved environment of the various body forms that characterize the developmental pathway of traits. Thus, what has evolved (including the existing developmental adaptations) may constrain what can evolve, or may establish certain evolutionary paths as more likely than others.

Because selection is the most important cause of evolution, and because it is the only evolutionary agent that can produce adaptations, the ultimate approach seeks to provide explanations for these seemingly purposefully designed biological traits of individuals in relation to the causal selective forces that produced them. Thus, the adaptationist approach focuses on how an adaptation contributed to successful reproduction of its bearers in the environments of evolutionary history. The challenge in applying an ultimate or evolutionary analysis is not to determine whether an adaptation is a product of selection; it is to determine the nature of the selective pressure that is responsible for the trait. That selective pressure will be apparent in the functional design of the adaptation.

By-Products of Selection

Not all aspects of living organisms are adaptations. Indeed, Williams (1966, pp. 4–5) emphasized that "adaptation is a special and onerous concept that should be used only where it is really necessary," and the evolutionists that Williams inspired have been well aware that a trait's mere existence does not mean that it was directly favored by natural selection. Nor is a demonstration that a trait or a character increases an individual's reproductive success sufficient evidence that the trait is an adaptation.

Not only may an increase in reproductive success be due to some evolutionarily novel aspect of the environment; an increase in reproductive success in evolutionary environments may be only a beneficial effect rather than an evolutionary function. To illustrate this point, Williams cited a fox walking through deep snow to a henhouse to catch a chicken, then following its own footprints on subsequent visits to the henhouse. This makes subsequent trips to the henhouse more energy efficient for the fox, thus potentially increasing its reproductive success. Following its own footprints back may well involve adaptations in the brain of the fox, but there is no known feature of the fox's feet that exhibits design by natural selection to pack snow. The fox's feet are clearly designed for walking and running, but they are not clearly designed for snow packing. Hence, even though snow may have been part of the past environments of foxes, there is no evidence that it acted as a sufficient selective pressure to design the feet of foxes for efficient snow packing. Snow packing and any associated reproductive success appear to be fortuitous effects of the structure of the fox's feet. That is, snow packing is not a function of any known aspect of the fox's feet. Symons (1979, p. 10) noted that "to say that a given beneficial effect of a character is the function, or a function, of that character is to say that the character was molded by natural selection to produce that effect." Williams (1966, p. 209) stated that "the demonstration of a benefit is neither necessary nor sufficient in the demonstration of function, although it may sometimes provide insight not otherwise obtainable," and that "it is both necessary and sufficient to show that the process [or trait] is designed to serve the function."[6]

As Williams emphasized, the concept of adaptation should be used only where really necessary; however, it is essential to *consider* the concept of adaptation in all cases of possible phenotypic design, because only then can it be determined if a trait has been designed by natural selection. Williams (ibid., p. 10) proposed that plausibly demonstrating design by natural selection requires showing that a trait accomplishes its alleged function with "sufficient precision, economy, and efficiency, etc."[7] Following Williams's criteria, Symons (1979, p. 11) stated that "[a] function can be distinguished from an incidental effect insofar as it is produced with sufficient precision, economy, and efficiency to rule out chance as an adequate explanation of its existence." Hence, according to the doctrine of parsimony, "if an effect can be explained adequately as the result of physical laws or as the fortuitous byproduct of an adaptation, it should not be called a function" (ibid.).

Similarly, drift and mutation can be ruled out as explanations of the evolutionary history of a trait when the trait shows evidence of functional design. Drift may apply only to traits that do not adversely affect reproductive success: if there are such effects, then selection will determine a trait's fate. Few traits meet the criterion of no cost to reproductive success; thus, as the biologists Richard Alexander (1979) and Richard Dawkins (1986) have explained, drift is a matter of interest primarily in the cases of phenotypic traits that do not attract adaptationists' attention in the first place.

Most mutations are deleterious and thus are in a balance with selection (selection lowering the frequency and mutation increasing it). Selection is stronger because mutation rates are very low. Thus, mutation, as an evolutionary cause for traits, may apply only to those traits that are only slightly above zero frequency in the population. Because selection is the most potent of the evolutionary agents, any explanation of the evolutionary history of a trait based on mutation or on drift must be fully reconciled with the potency of selection to bring about trait evolution.

Further evidence of adaptation may come from cross-species comparisons. First, "if related species [i.e., those sharing a recent common ancestral species] come to occupy different environments where they are subject to different selection pressures, then they should evolve new traits as adaptive mutations occur that confer a reproductive advantage under the new conditions" (Alcock 1993, p. 222). Variation among the beaks of different species of the finches Darwin found on the Galápagos Islands would be an example of such "divergent evolution." The beak types are different adaptations for eating different, species-typical foods (Weiner 1994). Second, if two distantly related species "have been subjected to similar selection pressures," they "should have independently evolved similar behavioral traits through *convergent evolution*—if the trait truly is an adaptation to that selection pressure" (Alcock 1993, p. 222). Convergent evolution is responsible for the similar shapes of fishes and marine mammals that have evolved by natural selection in the context of mobility in water.

Hence, the diversity of life has two major components: adaptations and the effects of adaptations. The latter are known as *by-products*. Adaptations are traits formed directly by selective pressures; by-products are traits formed indirectly by selective pressures.

In addition to snow packing by fox feet, another example of a by-product is the red color of human arterial blood (Symons 1987a,b). This trait did not arise because of selection in the context of blood-color variation among individuals. That is, redness of arterial blood did not cause individuals with arterial blood of that color to become more frequent in succeeding generations. Instead, selection acting in other contexts gave rise to the trait as an epiphenomenon of adaptations. Human arterial blood is red for two proximate reasons: the chemistry of oxygen and hemoglobin in blood, and human color vision. Hence, the ultimate causation of the color of blood lies in the selective pressures that produced the chemical composition of human blood and human color vision.

Another example of a by-product is the higher death rate of males relative to females among humans of all ages (Alexander 1979; Trivers 1985; Wilson and Daly 1985; Geary 1998). The higher male mortality is not an adaptation; it is an incidental effect of sex-specific adaptations. The adaptations are in males' and females' bodies, including their brains. For example, various traits motivate male humans, relative to female humans, to engage in riskier activities. The ultimate cause of these male adaptations is a human evolutionary history of stronger sexual selection acting on males than on females.[8]

When one is considering any feature of living things, whether evolution applies is never a question. The only legitimate question is how to apply evolutionary principles. This is the case for all human behaviors—even for such by-products as cosmetic surgery, the content of movies, legal systems, and fashion trends.

The crucial legitimate scientific debate about the evolutionary cause of human rape concerns whether rape is a result of rape-specific adaptation or a by-product of other adaptations. That is, does rape result from men's special-purpose psychology, and perhaps from associated non-psychological anatomy, designed by selection for rape, or is rape an incidental effect of special-purpose adaptation to circumstances other than rape? We two authors, having debated this question for more than a decade (Palmer 1991, 1992a,b; Thornhill and Thornhill 1992a,b), agree that it may eventually be answered by determining whether or not rape is the result of special-purpose psychological mechanisms that motivate and regulate men's pursuit of rape in itself. We also agree that enough now is known about the ultimate evolutionary causes of human rape that an evolutionary approach can contribute significantly to prevention of the act.

But how can an ultimate explanation of why men rape help prevent future rapes? The answer is that ultimate evolutionary explanations have unique power in both a theoretical and a practical sense. In terms of theory, only selection can account for the creation and the maintenance of adaptations. Even complete identification of all proximate causes of an adaptation could not explain the genesis and the persistence of that adaptation. However, an ultimate explanation of a biological phenomenon can account for all proximate causes influencing the phenomenon, whether the phenomenon is an adaptation or an incidental effect of an adaptation. Thus, ultimate explanations are more general in that they are more inclusive of causation. As a result, ultimate explanations have enormous practical potential: if evolution by individual selection is truly the general theory of life, it should lead to the best insights about proximate causes, and identifying proximate causes is the key to changing human behavior (e.g., eliminating rape).

That an ultimate evolutionary approach can serve as a guide for research into proximate causes has been shown repeatedly in investigations of nonhuman organisms. Indeed, this approach has revolutionized those investigations (Krebs and Davies 1993; Alcock 1997). It is also revolutionizing the study of human behavior (Alexander 1987; Wright 1994; Pinker 1997; Geary 1998; Buss 1999).

Evolutionary theory contributes to the study of proximate causation in two ways.

First, it leads to the discovery of new biological phenomena whose proximate causes are unknown. For example, the evolutionary psychologists Leda Cosmides and John Tooby (1992) have found that the human brain contains a mechanism designed specifically to detect cheating in social exchanges. The discovery of such a "cheater-detection" mechanism was the result of an understanding of the evolutionary concept of reciprocal altruism originally developed by the biologist Robert Trivers (1971). Similarly, evolutionary theory has led to the discovery of specific patterns of nepotism. This knowledge has resulted from studies directed by the fundamental evolutionary concept of kin selection: individuals perpetuate their genes not only by producing offspring but also by aiding relatives, including offspring (Hamilton 1963, 1964; Alexander 1987; Chagnon and Irons 1979; Betzig et al. 1988; Betzig 1997; Crawford and Krebs 1998). Relatives contain a high proportion of identical genes, and the closer the kinship relationship the higher the genetic similarity. What are the proximate cues by which individuals identify their relatives and distinguish categories of relatives? "Social learning" is the general answer (Alexander 1979; Palmer and Steadman 1997). Children are taught who their relatives are by their parents and their other relatives and through association with them during upbringing, and are encouraged by their adult relatives to be altruistic toward them (especially close kin). But what is the precise nature of the learning schedules involved in the ontogeny (development) of an individual's nepotistic behavior? This question would never have been asked had not evolutionists first successfully predicted the patterns of nepotistic behavior. After the social-learning aspects of nepotism are understood, the proximate physiological mechanisms in the brain that cause humans to feel closer to and more generous toward close relatives can be investigated. Also, we may someday know the locations of human genes (another category of proximate causation), which, in conjunction with the environment, construct proximate mechanisms of kin recognition and discriminative nepotism.

The second way in which evolutionary theory interacts with the identification of proximate causes is even more direct and important. Evolutionary theory can tell investigators what proximate mechanisms are most likely to be found, and therefore where any investigation of proximate causation should begin. For example, evolutionary theory has provided unique directions for investigations of child abuse, child neglect, and infanticide (Daly and Wilson 1988). Evolutionary predictions regarding parental investment have directed researchers to multiple proximate causes of child maltreatment: resources available for successfully rearing offspring; paternity certainty and genetic relatedness of parent to offspring generally; health, sex, and status of offspring; age of parent; birth order.[9]

The example of child abuse also demonstrates the ability of an evolutionary approach to identify the proximate causes of both adaptations and by-products. In this case, it is not child abuse or infanticide per se that was favored by selection in human evolutionary history. The adaptations concern what Daly and Wilson (1988) call "child-specific parental solicitude" or "discriminative parental solicitude," which evolved because they increased the number of surviving offspring in a parent's lifetime relative to parents who invested indiscriminately in children generally. These are species-wide psychological adaptations that cause some parents to show love to all their children more or less equally, or to love some children and neglect (or even abuse or kill) others. The power of an evolutionary approach in identifying these factors is illustrated by Daly and Wilson's observation (1995, p. 22) that "living with a stepparent has turned out to be the most powerful predictor of severe child abuse risk yet discovered, but two decades of intensive child abuse research conducted without the heuristic assistance of Darwinian insights never discovered it." We suggest that the evolutionary approach can make a similar contribution to the identification of the proximate causes of rape. Specifically, we suggest that an understanding of the evolved differences between male and female sexuality can lead to identification of the proximate causes of rape. Indeed, the ability of an ultimate evolutionary approach to direct research to the proximate causes of rape may be the key to lowering the frequency of rape.

Adaptations Are Functionally Specific

An understanding of the ultimate cause of adaptations can provide specific ways of preventing rape because adaptations are themselves specific.

In a paper titled "If we're all Darwinians, what's the fuss about?" Donald Symons (1987a) pointed out that the difference of opinion between traditional social scientists and the evolutionary anthropologists, biologists, and psychologists who were inspired by Williams's book *Adaptation and Natural Selection* does not concern whether or not the brain is designed by selection. The idea of psychological (brain) adaptation is almost certainly compelling to anyone who accepts that the rest of the human body has evolved by Darwinian selection. Indeed, the notion that the rest of the body could have been designed by selection without selection's simultaneously acting on the brain and the nervous system that control the body is absurd. To those who accept the notion of evolution, it is clear that the human brain must contain evolved structures that process environmental information in a manner that guides feelings and behavior toward ends that were adaptive in past human environments. Similarly, a moment's reflection on the evolution of the human opposable thumb—whose name implies both a structure and the movement (behavior) of that structure—should resolve any remaining controversy as to whether human physical behavior (muscle-induced motion) has evolved. All this means that the explanations of human behavior put forth by the social scientists who accept evolution (the vast majority) are implicitly evolutionary explanations. Hence, according to Symons (p. 126), "perhaps the central issue in psychology is whether the mechanisms of the mind are few, general, and simple, on the one hand, or numerous, specific, and complex, on the other." Symons goes on to say that "for all their differences, theories that purport to explain human affairs in terms of *learning*, *socialization*, or *culture*, and so on seem to have one thing in common: they assume that a few generalized brain/mind mechanisms of association or symbol manipulation underpin human action" (p. 139). We suggest that one reason that many social scientists have not learned evolutionary theory is that they have mistakenly assumed that adaptations are so general as to be of little significance.

Special-Purpose and General-Purpose Adaptations

Defined more precisely than above, adaptations are mechanisms that Darwinian selection "designed" because they provided solutions to environmental problems faced by ancestors (Williams 1966, 1992; Symons 1979; Thornhill 1990, 1997a). Providing these solutions is the "function" of adaptations (Williams 1966).

Although most people consider physical traits to be distinct from psychological (or mental) traits, this is a mistake. The brain, even if one calls it the psyche, is a physiological component of the body. In fact, the brain is the component of physiology and anatomy that controls the rest of physiology and anatomy via environmental information processing. Hence, when evolutionary psychologists speak of evolved "psychological mechanisms," they are actually postulating physiological mechanisms in the nervous system that, at the present stage of scientific knowledge, can only be inferred from patterns of behavior (Palmer 1991, 1992a,b).

Psychological mechanisms can be characterized as either *special-purpose* or *general-purpose* on the basis of the kind of information they process to accomplish their function. Information that is *domain-specific* (for example, that will help an individual acquire a proper diet or a mate with high reproductive potential) is, by definition, special-purpose. If the information processed to accomplish a goal is ecologically general, the mechanism is, by definition, general-purpose. Thus, we can imagine a general-purpose mechanism that evaluates a broad range of items (food items, potential mates, rocks) in terms of their quality.

Hypothetically, adaptations could range from very general to very specific. For example, a mechanism that used the same information to obtain a good diet and a mate with high reproductive potential would not be as general-purpose as a mechanism that used the same information to solve those problems and also the problem of finding safe places to sleep. On the other hand, finding a mate with high reproductive potential might involve a number of even more specific mechanisms. For example, among humans there seem to be separate, specific psychological mechanisms that have evolved to discriminate health, age-related cues, and parenting ability in a potential mate (Symons 1979, 1995; Thornhill and Møller 1997; Townsend 1998).[10]

Hence, what is at question is not whether psychological mechanisms are general-purpose or special-purpose; it is their degree of specificity. Many social scientists believe that humans possess only a few very general psychological mechanisms; evolutionary psychologists posit many very specific mechanisms. This evolutionary perspective is akin to many cognitive scientists' long-standing assumption of the modularity of mind (Gazzaniga 1995).

There are three reasons why evolutionary psychologists argue that the human brain must be composed of many specialized, domain-specific adaptations.

The first is that **the environmental problems our evolutionary ancestors faced were quite specific.** Since adaptations are solutions to these specific environmental problems that impinged on ancestors during evolutionary history, they should be equally specific. Selection should have led to special-purpose adaptations because such adaptations can better solve specialized problems.

Any environmental problem that is typically solved by organisms could be used to illustrate the issue of specificity. Vision, for example, may at first appear to present only the very general problem of viewing one's environment. However, "vision" and "environment" are actually general words for complex phenomena. "Vision" entails solving many specific problems: color, black and white, depth, edges, distance, available light, and so on. Which of these problems an organism solves, and in what manner, will depend on very specific variables in the environment in which the organism's ancestors lived. Hence, the eyes, brains, and nervous systems of various species respond only to certain colors, shapes, and movements, and these vary greatly among species in correspondence to the features of the environments that impinged on the past reproductive success of individuals of the various species. For example, some cells in the European toad's eye "respond most to long, thin objects that move horizontally across the toad's visual field," and this specific design "becomes clear if one imagines how they would respond to a nearby moving worm" (Alcock 1993, pp. 134, 135). Furthermore, an individual animal's environment often is specific not only to the species but also to the individual's age and sex. Vision stems from many specialized psychological adaptations, each designed to process specific environmental information.[11] An eye is a collection of many special-purpose psychological adaptations. Evolutionary psychologists expect the same to be true of an organism's other adaptations.

The second reason why human psychological adaptations are expected to be special-purpose is that **much of successful human behavior depends on environmental circumstances that are variable** (Symons 1987a).[12]

The existence of environmentally dependent behavioral flexibility is often mistakenly used by social scientists to argue *against* the existence of specialized brain structures. "Many writers seem to believe that behavioral flexibility somehow implies the existence of simple, amorphous mental structures," Symons (1987a, p. 127) notes. He continues: "There is a litany in the literature of anthropology that goes something like this: Human beings have no nature because the essence of the human adaptation is plasticity, which makes possible rapid behavioral adjustments to environmental variations. This litany, however, has the matter backwards: Extreme behavioral plasticity implies extreme mental complexity and stability; that is, an elaborate human nature. Behavioral plasticity for its own sake would be worse than useless, random variation suicide. During the course of evolutionary history the more plastic hominid behavior became the more complex the neural machinery must have become to channel this plasticity into adaptive action."

A *facultative* response to the environment (that is, a conditional response that depends on specific environmental variables) evolves when the environment changes within the lifetime of an individual in a way that significantly influences reproductive success. The capacity to learn is one such response. The human social environment is one of change, and the portion of human psychology that is involved with social learning is large. This is probably an evolutionary outcome of selection in the context of changing social conditions within the lifetimes of individuals, coupled with an inability to solve a learning task by experimentation or trial-and-error learning; under this scenario, social learning evolves (Humphrey 1980; Alexander 1989). However, learning will generate *maladaptive* behaviors (behaviors that decrease the reproductive success of the individual) unless special-purpose mental mechanisms guide and bias learning and behavior along paths that are adaptive.

We humans are social strategists par excellence (Wright 1994), and our social behavior is apparently unique in the degree of its plasticity. This unique behavioral plasticity requires not only that human psychology consists of many specialized mechanisms but also that it be much more diverse and complex in structure than the psychology of any other organism.

The third reason that human psychological adaptations are expected to be special-purpose rather than general-purpose is that **our knowledge of the functional design of non-psychological adaptations indicates that they are special-purpose.** The human body, for example, is not a single general-purpose adaptation; it is a bundle of innumerable specific adaptations designed to solve specific challenges to reproduction in past environments. Indeed, those who accept the reality of evolution realize that species-specific non-psychological adaptations are what allow biologists to distinguish species morphologically, physiologically, and developmentally. If adaptations were general-purpose, differences among species (including differences in behavior) would not exist, and thus the discipline of taxonomy (the classification of organisms) would not exist. It is also sex-specific adaptations, psychological and otherwise, that allow researchers to describe sex differences, and it is age-specific adaptations, psychological and otherwise, that make the field of developmental biology possible.

Many social scientists apparently fail to realize that it is species-specific *psychological* adaptations that allow biologists to distinguish species *behaviorally*. Not only is it unreasonable to think that the human psyche will be an exception to the general pattern of specific adaptations; there is increasing evidence from behavioral studies and from neuroscience that the human psyche is composed of adaptations that process specialized information.

In 1989 the cognitive neuroscientist Michael Gazzaniga reviewed the evidence that aspects of human cognition are structurally and functionally organized into discrete units ("modules") that interact to produce mental activity. Gazzaniga summarized his review as follows: ". . . when considering the various observations reported here, it is important to keep in mind the evolutionary history of our species. Over the course of this evolution efficient systems have been selected for handling critical environmental challenges. In this light, it is no wonder there are specialized systems (modules) that are active in carrying out specific and important assignments." (1989, p. 951) As is evident from this summary, Gazzaniga had been led by empirical evidence to the conclusion that the human psyche is made up of many specialized adaptations.

Of course, to demonstrate the implausibility of the assumption that there are only a few very general psychological adaptations is not to demonstrate the existence of very specialized adaptations. Similarly, the existence of specialized adaptations in the frog brain is not evidence that similar specialized adaptations exist in the human brain. But evidence of specialized adaptations in the human brain is abundant. Symons (1987b, 1992), Cosmides and Tooby (1987, 1989), Barkow et al. (1992), Buss (1994, 1999), Gazzaniga (1995), Pinker (1997), and many others have amassed human behavioral evidence that the specific nature of ecological problems applies to environmental information-processing problems as much as it applies to other related problems, and thus that human psychological mechanisms appear to be domain-specific in function.[13]

Although evolutionists debate the exact degree of specificity of the psychological mechanisms of the human brain (Symons 1987b, 1992; Alexander 1990; Turke 1990), essentially all of them are in agreement that the brain is much more specialized than is implied by a certain class of social scientists. As the evolutionary anthropologist Paul Turke (1990, p. 319) notes, "with the exception of some outdated behaviorists, . . . we all have been working towards understanding the nature of the more or less specific mechanisms that constitute the human psyche."

Biology, Learning, and Ontogeny

Social scientists commonly assert that cultural learning is not biological, evidently because they inaccurately equate "biological" with "genetic." In reality, every aspect of every living thing is, by definition, biological, and everything biological is a result of interaction between genes and environmental factors. Without this understanding, it is not possible to understand how domain-specific adaptations of the human brain develop and how they are involved in learning.

Even an individual cell—the most fundamental building block of any larger organism—is a product of genes *and* certain aspects of the environment (e.g., various chemicals). Certain changes in either the genes or the environment change a cell (and may even end its existence). As an organism continues to develop, genes will create new cells only when they interact with certain additional environmental triggers, and differences in the developmental environment will produce a variety of cells (muscle cells, nerve cells, and so on). This constant intertwining of genetic and environmental factors continues throughout the life of the organism. The environmental factors include not only a multitude of things external to the individual (oxygen, water, nutrients, other individuals, and so on) but also the environment within the developing individual (e.g., other cells, tissues, organs). And these cells, tissues, and organs themselves are products of their own gene-environment interactions.

The interaction of genes and environment in development is too intimate to be separated into "genes" and "environment." Not only is it meaningless to suggest that any trait of an individual is environmentally or genetically "determined"; it is not even valid to talk of a trait as "primarily" genetic or environmental. However, since "biological" actually means "of or pertaining to life," it is quite valid to claim that any phenotypic trait of an organism is biologically, or evolutionarily, determined (Daly and Wilson 1983, chapter 10; Oyama 1985). Genes per se are not evaluated by selection. Instead, it is the interaction of genes and environment that selection evaluates. When a given interaction produces a trait that promotes individual reproduction more than an alternative trait created by a different gene-environment interaction, the genetic underpinning of the reproductively superior trait increases in frequency in the population. When selection acts in a directional manner over a long period of time, gene frequencies change, gene-environment interactions change, and new adaptations spread. Adaptations, then, as Tooby and Cosmides (1990a) and others have emphasized, are manifestations of *evolved* gene-environment interactions. Thus, the environmental and the genetic causes acting during development are not only equally important and inseparable; in addition, they are specific and non-arbitrary. Both the environmental and the genetic causes reflect evolutionary history, and equally so.

Biological or evolutionary determinism is not equivalent to biological inevitability. Indeed, the accretion of scientific knowledge about how traits develop, with equal causal input from genes and from environment, makes it more likely that traits can be altered by changing one or more of their developmental causes.

The degree to which differences between individuals are due to differences in genes—known as *heritability*—is expressed as the proportion of the variation among individuals with regard to a certain trait that is attributable to genetic rather than environmental variation (Falconer 1981). For example, difference between individual humans in height has a heritability index as high as 0.9 in some human populations (Bodmer and Cavalli-Sforza 1976). This means that about 90 percent of the difference in height between individuals is due to genetic differences, and about 10 percent to differences in environment (nutrition, disease, etc.). However, the height of any individual is the result of an inseparable interaction of genes and environmental factors. Hence, height (like all aspects of living things) is "biologically determined," because it is the product of both genetic and environmental factors.

That heritability is a very different concept than inheritance is evident from the fact that inheritance occurs in the absence of heritability. For example, although two hands are normally inherited from one's parents, hand number is not a heritable trait—that is, there is essentially no genetic variance underlying hand number. In times past, hand number in humans was under strong selection, and that greatly reduced genetic variation affecting the development of this trait. In other words, the genes that encode for two hands are virtually fixed in humans.

Thus, Michael Crichton's *Jurassic Park* is truly fictional. Even if someone were to obtain the fossilized DNA of extinct dinosaurs, transferring those genes to an iguana egg would not yield a dinosaur. The genes of a *Tyrannosaurus rex* could express themselves adaptively only in the environment of a *T. rex* egg, then in that of a *T. rex* embryo, fetus, hatchling, and adult—an environment that is as extinct as *T. rex* itself.

Learning

Social scientists often treat learning as a distinctive—indeed, even a nonbiological—phenomenon. They also view it as a complete, or an essentially complete, explanation of behavior. In fact, however, learning is only a specific type of gene-environment biological interaction. It is, therefore, one type of proximate cause—to be more specific, one type of developmental cause.

Both learned and non-learned "innate" behaviors are products of gene-environment interactions. Either requires interaction between genes and a vast number of things in the environment. These two types of behaviors are distinguished only by whether or not one specific *identified* aspect of the environment is among the environmental factors that must be present for the behavior to occur.

We call a behavior "learned" when we have identified a specific experiential factor as necessary for its occurrence. For example, it is because we have identified that one must pick up a bow and shoot an arrow several times before one is likely to hit a target that we call archery a learned behavior. Similarly, "innate" behaviors require previous interactions with many specific environmental factors during the development of the organism. The word 'innate' only connotes that certain environmental exposures are *not* necessary in order for the behavior to occur. For example, the sucking behavior of newborn infants is often called innate because it doesn't require the specific previous environmental influence of exposure to a nipple. But this use of 'innate' overlooks the fact that the behavior requires the presence of many other environmental factors.

Rather than implying that *no* environmental factors are necessary, "innate" actually implies that specific environmental factors necessary for the behavior to occur have not been identified. Hence, it connotes behaviors and other traits for which *particular* experiences can be ruled out as developmental causes, but not *all* environmental causes. Conversely, "learned" implies only that specific identified environmental factors necessary to the occurrence of the behavior have been identified, not that such environmental factors are sufficient for the behavior to occur.

The modern view of development means that psychological adaptations, including those that affect human sexuality, have been designed by selection during our evolution to process specific, non-arbitrary information in the environment. Such design is the case whether a psychological adaptation requires experiences with environmental stimuli commonly referred to as "learning" or whether it is influenced only by other experiences during ontogeny that do not fit standard definitions of learning (Symons 1979, pp. 17–21). Individuals whose psychological mechanisms did not guide behavior, feelings, development, hormone release, and so on adaptively in human evolutionary history are no one's evolutionary ancestors. Adaptive psychological traits of individuals that increased in frequency during human evolution had one essential property that made them, and not alternative traits, successful in withstanding selection: They helped individuals reproduce successfully in ancestral environments because they contributed to the solution of specific environmental problems.

Perception and processing of arbitrary environmental information by psychological features will lead to psychological changes and behavioral effects that provide ineffective responses to environmental challenges that cause selection. Thus, each psychological adaptation (and each non-psychological adaptation) has evolved because of a precise, specific, *non-arbitrary* relationship between genes and environment. Learning abilities and underlying psychological mechanisms cannot be isolated from genes, adaptation, and our evolutionary past.

Culture

Is the socially learned behavior known as culture still biological and subject to the only general biological theory—evolution by selection? A common justification for rejecting evolutionary explanations of human behavior is that it is not, and that hence it requires an entirely different approach. This view was expressed recently by the feminist biologist Anne Fausto-Sterling (1997, p. 47): "I have found it useful to try to separate discussions of sociobiological approaches to the study of animal behavior from the application of such approaches to human behavior. I do this, not because I believe in a special, non-evolutionary creation for humans. Rather, I think that the evolution of culture has enormously complicated the task of understanding human behavior and development."

Although culture certainly hasn't simplified the task of understanding human behavior and development, has it really removed some human behaviors from the realm of biology and evolutionary explanation? Are some human behaviors biological and others not? The feminist biologist Victoria Sork (1997, p. 89) refers to "gender differences in human society—some of which are biologically based, and some of which are culturally based." The confusion here is the same one we pointed out in our discussion of heritability. Yes, some differences in behavior between individuals could be due entirely to cultural influences that have affected their behavior. But that is very different from saying that an individual's culturally influenced behavior is due entirely to environmental causes and hence is not biological. An individual's cultural behavior is still a product of gene-environment interactions. And the individual can learn nothing without underlying adaptation for learning.

Most social scientists use the word 'culture' when referring to socially learned behavior (Flinn 1997). Although culture is often asserted to involve mental states, and sometimes asserted to involve only mental states, we know that we are dealing with culture only when we observe certain kinds of behavior or their consequences. The realization that culture is behavior places it clearly within the realm of biology, and hence within the explanatory realm of natural selection.

That culture is socially learned behavior means only that the causes of the behavior *include*, not that they are limited to, learning experiences involving other human beings (Steadman and Palmer 1995). Just as some people use the word 'learned' to refer to the subset of behavior for which we have identified a specific necessary environmental factor, some use the word 'cultural' to refer to the subset of learned behavior for which we have identified that a specific necessary environmental factor is another person.

Speaking a language, for example, is clearly a cultural behavior, because the environmental influences leading to its occurrence *include* social learning. It does not follow, however, that "cultural evolution can facilitate the transmission of behaviors from one generation to the next as well as within a generation without any genetic basis" (Sork 1997, p. 109). Although learning experiences involving members of the same species are necessary for language acquisition, they are far from sufficient for it. Among the other necessary proximate precursors to speaking a language is a set of specialized brain structures forming at particular stages of development that are themselves the ultimate product of a long history of natural selection and the proximate product of complex gene-environment interactions during ontogeny (Pinker 1994). Hence, although language is cultural, it is still just as biological, and just as subject to evolutionary influences, as the human eye.

The parent-offspring resemblance that has typified language, and until a few thousand years ago nearly all other aspects of culture, is typically referred to as *tradition*. Cultural traditions result when both environmental and genetic influences on the trait are repeated across generations. Specific genes and specific environments interact during development to produce adaptations in young humans that enable them to learn a language from others. Genes are passed in the gametes of parents. Male and female gametes unite to form the zygote. The genes of the developing individual interact with the environment—that is, with everything external to the specific genes being expressed during development: cytoplasm, nourishment, the developing individual itself, other genes. The gene-environment interaction results in nervous-system adaptations that make possible the perception and processing of information. Genetic and environmental influences also construct the emotional and cognitive adaptations[14] of the brain, including those involved in the copying of behavior and the highly specialized mechanisms designed to copy language. If the social learning (copying) involves English in parent and offspring generations, there will be parent-offspring resemblance and the behavior of speaking English can be said to have been inherited.

As evidence that *both* genetic and environmental influences must be transmitted in order for language to be inherited, consider what factors eliminate the inheritance of language. Suppose that a young child of English-speaking parents is adopted and then raised in an environment in which only French is spoken. In this case, the speaking of English is not inherited, because the language spoken in the environment during certain stages of development was not repeated in the environment of the offspring. Now suppose that a child raised by its parents in an environment in which only English is spoken does not learn English despite the opportunity. In the latter case, the child may have received genes not expressed in the parents (recessive genes)—for example, the child may be deaf as a result of such genes.

When inheritance is properly considered as a phenotypic phenomenon caused by both genetic and environmental causes, there can be no confusion about how cultural behavior is inherited.[15] There is no fundamental difference in the mechanisms of inheritance of cultural and non-cultural behavior, nor is there a difference between the mechanisms of inheritance of cultural behavior and the mechanisms of inheritance of physiology and morphology. Inheritance occurs—like begets like, traits breed true—when and only when both genetic and environmental influences are repeated between generations.

This approach provides an answer to the question about the relationship between culture and biology that has dominated much of the history of social science (Freeman 1983; Brown 1991). Claims that cultural inheritance is independent of biological inheritance, whether made by nonevolutionary social scientists or by evolutionary biologists (Dawkins 1976; Boyd and Richerson 1978, 1985; Pulliam and Dunford 1980), are erroneous. Culture is not the "superorganic" force that some social scientists have claimed it to be. Nor, as the philosopher Daniel Dennett (1995) has pointed out, does culture consist of ideas (also called *memes*) that parasitize minds independent of psychological (biological) adaptation, as certain biologists have claimed. Such claims are simply inconsistent with modern knowledge of how inheritance and development work.

Gregor Mendel discovered the role of genes in inheritance, but of course he did not discover inheritance itself. That like begets like was known long before Mendel. Parent-offspring resemblance in socially learned behavior requires psychological mechanisms—mechanisms that are produced by gene-environment interaction during development. Innate behavior, learned behavior, and cultural behavior are all products of brains. Brains are products of gene-environment interactions. Gene-environment interactions are subject to natural selection.[16]

However, a given cultural behavior cannot be automatically assumed to increase current reproductive success, nor can it be assumed to have been designed by natural selection (even though our *capacity* for culture clearly was). Cultural behavior, like all behavior, should be expected to show evidence of adaptation, and thus direct selection for the behavior, only to the extent that both the genetic and the environmental influences on that behavior have been replicated across generations for the long periods of time needed for effective selection. Since we use the word 'tradition' to refer to such enduring cultural behavior, it is to the extent that cultural behavior is traditional that it is expected to show evidence of selectionist design.

The greater the number of generations in which a cultural behavior has been replicated, the greater is the probability of evidence of design. At one extreme are certain cultural behaviors, such as an individual's adoption of a new hairstyle, that show no evidence of design by natural selection. Although a hairstyle is a by-product of numerous underlying psychological adaptations (perhaps concerning status, mate preferences, and/or visual acuity), a particular new hairstyle in itself cannot be considered an adaptation. At the other extreme are cultural behaviors that may have been copied for hundreds or even thousands of generations, thus implying the replication of both the genes involved and the environmental influence of the behavior of other individuals in each generation. In addition to language, such extremely traditional cultural behaviors include aspects of child care (feeding and caring), systems of kinship identification (kin terms, descent names, clan markings), techniques for manufacturing stone tools, hunting strategies, religious rituals, mating practices, and systems of punishment.

Although there is much variation in traditions, the universal presence of kin terms, religious rituals, and languages (Brown 1991; Steadman and Palmer 1995; Steadman et al. 1996) suggests the role of species-typical adaptations in all these behaviors. Thus, although culture can change much faster than adaptations (as a result of changes in the environmental factors that contribute to the cultural behavior), cultural traditions and their underlying psychological mechanisms also show obvious signs of adaptation. For example, a multitude of new words may enter a language during a short period of time, while language itself remains a highly adaptive, evolved vehicle for communication.

Like any aspect of phenotype, any cultural behavior—whether designed by natural selection or merely a by-product of other adaptations—may be currently non-adaptive, and even maladaptive, as a result of environmental influences that are novel with respect to the historical environments in which the mechanisms responsible for the cultural behavior evolved. Just as some of the pronghorn antelope's social behaviors are not currently adaptive but are adaptations to extinct predators, many human adaptations, both behavioral and physiological, are not currently adaptive.

Even when cultural change and adaptation do not coincide, applying evolutionary principles to human behavior is still valid. Human psychological mechanisms of social learning are still the products of a long history of selection, and they still affect the cultural behavior of individuals—even when they produce novel, non-traditional cultural behavior. The cultural behavior of individuals is *never* independent of the human evolutionary history of selection for individual reproductive success.

Consciousness

Although there is considerable debate over the exact function of consciousness, and some debate over whether it is an adaptation or a by-product, there is no scientific reason for assuming that consciousness is anything other than an aspect of our evolved biology.

One hypothesis, proposed by the evolutionary psychologist Nicholas Humphrey (1980) and by the biologist Richard Alexander (1989), is that conscious awareness permits quick adaptive adjustments of social striving based on the perception of how well one is doing in social competition. The *consciousness adaptation* stores information on how others view one and helps one build and evaluate alternative scenarios that may promote one's success in the social arena. Humphrey and Alexander suggest that the most important aspect of the design of consciousness may be its usefulness for solving social problems that resemble, but differ slightly from, the social problems that were consistently faced in ancestral environments. Such new social problems are often evolutionarily unique combinations of the non-unique social variables that have been repeatedly encountered in human evolutionary history. For example, having to compete with other individuals of a certain age and sex for a resource such as mates is a task that was faced innumerable times by our ancestors, but the exact combination of the sexual variables we face on any given day may be unique. If consciousness is the result of our evolutionary history, it is almost certainly composed of special-purpose mechanisms for information processing (Alexander 1989, 1990; Turke 1990) that aid in solving problems composed of evolutionarily unique combinations of variables.

As an illustration of how consciousness processes information, consider an argument between two academics: A, who understands evolutionary biology, and B, who doesn't. Assume that each wants to win the argument. Suppose A is being interviewed by B for a job in a traditional social science department. Suppose that during the interview B argues that, because human behavior is cultural, it is independent of biology (and, therefore, A should not be hired). Knowing that many people erroneously equate biology with genetics, A launches into a long speech about gene-environment interactions, ontogeny, and psychological adaptations and looks for signs of comprehension from B—nods, affirmative grunts, perhaps even a smile. When no such responses are forthcoming, the now-perspiring A consciously searches her memory for statements that have produced signs of understanding in the past and hence may do so again. Detailed descriptions of the visual systems of frogs and references to the statistics on child abuse by stepparents spill forth. At one point, A even attempts to draw on a napkin a diagram illustrating an optical illusion that involves the edges of black squares. (See note 13.)

The point we wish to make is that this particular line of disagreement was not a part of the environment of human evolutionary history. It is evolutionarily novel. In this case, however, both individuals process *specific* information in their social-striving mechanisms. Each uses specific information about how the other has responded to his or her comments to decide on the next line of argument to use. Furthermore, they both use specific psychological procedures to construct specific arguments that relate this feedback to information that they learned in the course of their training in their respective fields. In this way, consciousness adaptations generate many kinds of secondary mechanisms or procedures that are used in social striving. Those mechanisms or procedures involve specific information—for example, that B sees "biological" and "cultural" as alternatives, but A doesn't. Furthermore, both participants process some of the same specific information. Their consciousness adaptations can be viewed as rules of conscious striving—and here "rules" means *specified procedures*, which implies special-purpose mechanisms. For example, the willingness of each individual to engage in and continue in the argument will depend on the perceived benefit to each party. This aspect alone, however, requires the processing of detailed, specific information, which typically differs considerably between individuals. This example illustrates that the psychological phenomena surrounding consciousness can be viewed as depending on specific information, and therefore as based on special-purpose mechanisms.

2

The Evolution of Sex Differences

Harbor seals are monogamous, and the males and the females are nearly equal in size. In contrast, male elephant seals are much heavier and longer than females, and a male elephant seal may inseminate as many as 100 females. Furthermore, the pronounced differences between the brains of male and female elephant seals produce vastly different sex-specific behavior patterns throughout their lives. For example, there are male-female differences in diet and in migration patterns. Perhaps the most striking sex-specific behavior among elephant seals, though, is the violent physical confrontations between males during the mating season.

The male-female differences in the brains, other body parts, and behavior of elephant seals are attributable to the simple fact that males and females in ancestral populations faced very different obstacles to reproduction. Hence, over thousands of generations, Darwinian selection favored different adaptations in males and females.

To understand why selection produced such different male-female adaptations in elephant seals but lesser differences between the adaptations of male and female harbor seals requires an understanding of what Darwin called sexual selection: the selection of traits that increase the quantity and/or the quality of an individual's mates rather than increasing the individual's ability to survive. Not incidentally, an understanding of sexual selection is also necessary to an understanding of the differences between the adaptations of male and female humans, and thus to a complete understanding of rape.

Sexual Selection in Humans

If female ancestors faced different environmental obstacles to reproduction than were faced by male ancestors in human evolutionary history, natural selection and sexual selection will have formed different adaptations in females and males.

It is obvious that men and women have evolved different *physical* adaptations. For example, the fact that women have functional breasts implies that female ancestors fed their infants with breast milk, and the greater upper-body strength of males implies physical competition among male ancestors. Although such evolutionary explanations of physical differences are relatively uncontroversial, many social scientists appear to be unaware that the examples just described are also evolved *behavioral* differences. Functional breasts would not have evolved without the simultaneous evolution of behavior patterns involving the placing of an infant to the breast (and the behavior pattern of sucking in the infant). Greater upper-body muscle mass in males would not have evolved without the simultaneous evolution of certain movements of those muscles (e.g., punching, shoving, grabbing). Furthermore, the evolution of these behavioral patterns implies psychological adaptations, both cognitive and emotional, to guide those behaviors. Acknowledging the evolution of physical (evidently referring to parts of the body other than the brain) sex differences while denying the evolution of the accompanying behavioral and psychological sex differences is not scientifically tenable.

Life Effort

To understand why human females and males have evolved different psychological and behavioral attributes, it is helpful to examine the evolutionary concept of *life effort*, defined technically as the total time, energy, and risk expended by an individual over its entire life span.

Although all the activities of an individual organism over its life span may influence its reproductive success, biologists often conceptually divide activities into reproductive effort and survival effort.[1] *Reproductive effort* (or *reproductive investment*) refers only to risks, structures, and activities that are directly related to reproduction; *survival effort* refers to all activities, bodily structures, and risks taken in association with the survival, maintenance, and growth of the individual. The components of survival effort are conceptually distinct from those of reproductive effort in that their effects on reproductive success are less direct. Typically, reproductive effort and survival effort both require investments of time, energy, and risk taking. That is, what an organism allocates to one form of effort cannot be used to promote the other.

If natural selection has equipped organisms with adaptations (i.e., traits that evolved because they increased reproduction more than alternative traits did), why don't organisms devote all their effort directly to reproductive effort? The answer is that sometimes reproductive success is promoted by growing larger, by living longer, by learning complex skills, and/or by teaching offspring. Complex social skills can be acquired through the social learning processes of imitation and instruction. The types of skills that are generally acquired through these processes of survival effort increased the reproductive success of our ancestors when they reached adulthood. Social learning after adulthood has been reached also represents this form of effort.

One subcategory of reproductive effort is investments related to producing offspring, including energy and time expended on acquiring mates. Many human emotional, cognitive, and motivational mental mechanisms fall into this category (referred to as *mating effort*) because they promote successful courtship and the maintenance of sexual relationships. But reproductive effort also includes the effort that goes into aiding offspring, grandchildren, siblings, nieces, nephews, and cousins.

Sex Differences

Darwin proposed that sexually selected traits either gave a male advantages in competition with other males for sexual access to females or increased a male's likelihood of being chosen as a mate by a female. The pioneering evolutionary theorist Robert Trivers (1972) provided the basic theory of what governs the extent of this sexual selection.[2] Trivers proposed that the effects of sexual selection are largely determined by the relative parental effort of the sexes in offspring production.

The reasoning that underlies Trivers's theory goes as follows: Because a population is a collection of interbreeding individuals, the parental effort of all the individuals of one sex is potentially accessible to each member of the opposite sex. Thus, parental effort will be the object of all competition among members of one sex for the opposite sex. Males will compete with other males to gain access to the parental effort of females, and females will compete with other females to gain access to the parental effort of males.

But how intense is the competition among individuals of either sex? Insofar as the intensity of competition hinges on how much can be won, the level of competition in one sex depends on the amount of parental effort provided by members of the opposite sex. Specifically, the level of competition in the sex with less parental effort will increase in proportion with the amount of parental effort provided by members of the other sex. If, say, males and females provide approximately equal parental effort, the competition among individuals of each sex will be about equal and of only moderate intensity. If the parental effort of females greatly exceeds that of males (as is often the case), there will be more competition among males for access to females' parental effort than there is competition among females for males' parental effort, and it will be much more intense. If the parental investment of males exceeds that of females, the reverse will hold.

In the terminology of evolutionary biology, parental effort is a *resource* that an individual can *possess* and that other individuals *desire*. (As used by evolutionary biologists, these terms do not mean that individuals need be consciously aware of the reproductive consequences of their emotions or actions.) Indeed, from an evolutionary perspective, parental effort is the essential resource, because it determines how many offspring there will be and their likelihood of surviving. The usual way of obtaining this resource from another individual is through sexual copulation, in which one individual's parental investment is, in a sense, "taken" by another and used by that individual to produce its offspring. This is why, if the amount of parental investment is unequal between the sexes, the sex that makes the greater parental investment becomes a limited resource for the other sex. Individuals of the desired sex are then in a position to choose mates, while individuals of the other sex must compete to get chosen.

In the majority of species, females exceed males in parental effort. Thus, across species, as the ratio of female to male parental effort increases, there should be a corresponding and direct increase in the extent of competition among males for access to females, while females should be increasingly choosy about their partners. Trivers's theory has been very successful in predicting the degree of sexual differences across animal species.[3]

The initial difference in parental investment—the difference in size between the sperm and the egg—has strongly biased selection to favor other adaptations that reinforce it. In many species, males direct all or most of their energy into trying to copulate, and females provide all the parental care. The result of this is that each male's role in offspring production (and hence his reproductive success) is limited entirely by his access to females, because females provide the parental effort upon which the survival of offspring depends. The number of successful offspring produced by females depends on effective expenditure of their finite parental effort, and this, coupled with the ready availability of many competing males, provides the basis for the evolution of female choice of a quality mate and of an appropriate time and place for the expenditure.

In species in which males engage only or primarily in mating effort, sexual selection on males has been maximally strong during evolutionary history, and thus the sexes show the extremes of evolved sexual differences, males often being larger, more colorful, and/or more pugnacious than females. But it is almost always the male sex that typically engages in the most sexually competitive activity. Males usually fight among themselves for females or for resources important to females. Males take the initiative in courtship. Males engage in risky activities in order to locate and to impress females. In general, whereas males often behave as if any female of their species (and sometimes females of other species, and sometimes inanimate objects) is a suitable mate and strive to encounter many mates, females act as if only certain males in the population are appropriate mates.

Trivers's theory of sex differences obtains its strongest support from the existence of a few species, including some fishes, frogs, birds, and insects, that show "sex-role reversal," females being more sexually competitive than males and males choosier about mates than females. Sex-role reversal is largely restricted to the very few species in which males provide more parental effort than females. For example, among pipefish (*Syngnathus typhle*), males supply nutrients and oxygen to the fertilized eggs for several weeks, and males favor large, ornamented females over small, plain ones

(Rosenqvist 1990). Also in accordance with Trivers's theory is the fact that differences in sexual behavior are relatively small in species with relatively small differences in parental effort between males and females.

In the few sexually reproducing species in which females offer less parental investment than males (known to biologists as *polyandrous* species), there is greater variation in offspring production among females than among males. This is a result of stronger sexual selection on females than on males. Species behave monogamously when the parental investment and the variation in offspring production of the sexes is equal (Gowaty and Mock 1985). In *polygyny* (where a portion of the males produce offspring with multiple mates), fewer males than females contribute genetically to each generation; this results in greater variation in offspring production among males than among females, since males have less parental investment and experience stronger sexual selection.

Many bird species have relatively little sex difference in parental investment. In these species, both sexes choose mates, compete for mates (to a degree, at least), and participate in parental activity. However, even in species with relatively similar parental efforts by the two sexes, males tend to show less parental effort, and this leads to greater sexual competition among males and more choosiness by females. The reason males show less parental effort in monogamous, biparental species is that the male in such species still allocates more effort to mating effort than the female does; this has evolved because the *minimum* reproductive effort required for successful offspring production by a male is very small relative to that required by a female. In humans, although the parental effort of males is sometimes similar to that of females, the difference in the minimum parental effort required to produce an offspring is enormous. Simply consider the difference in the potential number of offspring produced by a human male and the potential number produced by a human female (Trivers 1972; Symons 1979). This disparity is critical to an understanding of human sex differences.

Polygyny in Human Evolutionary History

Humans are mildly polygynous (Alexander 1979; Daly and Wilson 1983; Geary 1998). The differences in typical parental effort between the sexes are small relative to those in most mammals, thanks to the large amount of parental effort often exhibited by human males. But in our species, as in most others, males may successfully reproduce by expending only a very small amount of time and energy (only as much as is needed to produce an ejaculate and place it in a female's vagina). In contrast, the minimum effort required for a woman to reproduce successfully includes the vastly greater amounts of time, energy, and risk taking associated with nine months of pregnancy, childbirth, nursing, and typically many years of child care. Since both sexes may achieve the same benefit from a copulation (i.e., an offspring), the difference in the minimum cost of copulation to males and females is predicted to have led to selection favoring more mating effort by men than by women. Relative to women, men can produce more offspring by frequent sexual encounters with different partners; thus, males would be predicted to allocate more reproductive effort to mating effort than females do. This generates a number of testable predictions. One is that many ancestors of current humans were successful male polygynists. Many types of evidence confirm this prediction.

There should be no doubt that humans have been polygynous throughout evolutionary history, with greater sexual competition among males than among females. The ethnographic record shows that from 80 to 85 percent of human societies have allowed harem polygyny (Daly and Wilson 1983; Betzig 1986). The most overwhelming evidence, however, is comparative. In mammals with a history of greater sexual selection on males than on females, evolutionary theory predicts the following[4]:

1. Males will be larger than females (Darwin 1872).[5]
2. More males than females will be conceived and born (Alexander et al. 1979).
3. Males will die younger as a result of physiological malfunction than females (Williams 1957; Hamilton 1966).
4. Males will engage in more risky activities in the context of acquiring mates than females (Darwin 1874; Trivers 1972; Daly and Wilson 1983).
5. Males will have higher mortality than females as a result of external causes, such as combat, disease, and accidents (Trivers 1972; Daly and Wilson 1983).
6. Males will exhibit more general aggression than females (Darwin 1874).[6]

7. More often than females, males will engage in escalating violent aggression that leads to injury and even death (Darwin 1874; Clutton-Brock et al. 1982; Daly and Wilson 1988).
8. Pre-adult males will engage in more competitive and aggressive play than pre-adult females (Symons 1978; Alexander 1987).
9. Males will be less discriminating about and more eager to copulate with females than vice-versa (Darwin 1874; Williams 1966; Trivers 1972).

In this book, we will focus on the evidence that humans meet the last of the above predictions and on the relationship between this fact and the causes of human rape. Hence, we will not go through the derivations of the other predictions, nor will we discuss the vast evidence supporting these predictions (all of which are applicable to humans).[7]

That human evolutionary history has been a history of greater variation in offspring production by males than by females (i.e., polygyny) is known as assuredly as anything can be known in science. There is no scientific justification for continuing to debate it. That it continues to be debated testifies to the lack of knowledge about evolutionary principles among many social scientists. Thornhill and Thornhill (1983, p. 138) wrote that "humans have morphological, developmental, sex ratio, mortality, senescence, parental and general behavioral correlates of an evolutionary history of polygyny shown by other polygynous mammals." Because a human evolutionary history involving greater sexual competition among males than among females is a necessary component of evolutionary hypotheses of human rape, this statement was criticized by two opponents of the evolutionary approach (Tobach and Sunday 1985, p. 132) on the grounds that "the suggestion that human evolutionary history demonstrates polygyny is purely speculative; there are no data concerning early hominid mating patterns" and "correlation does not imply causality." These criticisms reveal the critics' lack of familiarity with evolution or with the comparative method that biologists use to study historical causation.

In reality, the data mentioned above arise from the question "What would one expect to see in behavior, morphology, physiology and development of current humans if human evolutionary history involved polygyny?" Consider, for example, the nine human sex differences just listed. These are the same sex differences seen in other mammals with polygynous evolutionary backgrounds. The human pattern is consistent with, or is correlated with, the patterns in polygynous mammals in general.

Tobach and Sunday (ibid.) also claim that "correlation does not imply causality." Indeed, one learns in a basic course on statistics that statistical correlation does not *necessarily* imply causality. To show that a correlation does imply causality, one must make further comparisons that control potentially confounding variables. In the case of the specific correlation criticized by Tobach and Sunday, it is informative to look at all comparative data, not just data on mammals. Sex differences 3–7 and 9 listed above are universally associated with polygynous species.[8] This means that the aspect of evolutionary theory dealing with sex differences has predicted successfully (i.e., the predictions have led to discoveries when tested) that sex differences 3–7 and 9 occur in species of all taxonomic groups whose evolutionary background was polygynous—and the more polygynous, the greater the magnitude of each difference. In regard to mammals, evolutionary theory argues that it is appropriate to add differences 1, 2, and 8. (Difference 8 appears to apply to many polygynous birds, too.) This kind of correlational evidence is very powerful in addressing causation. Regardless of taxonomic group, body size of species, adult life span, diet, habitat, and other factors, polygyny is associated with the sex differences. Because of the nature and the number of the comparisons that allow repeated demonstrations of evolutionary divergence of species within a taxon (a group of species descended from a common recent ancestral species) and of adaptive convergence of species in different taxa, there can be no reasonable doubt that creatures with sex differences 1–9 came from evolutionary backgrounds in which sexual selection was greater on males than on females.

The human evolutionary history of polygyny presented males and females with very different environmental challenges to reproduction. Hence, present-day men and women, as descendants of the members of ancestral populations who responded to those challenges most successfully, have very different psychological adaptations. We will discuss these psychological adaptations in terms of male *sexual preferences* and female *mate choice*. The reason we use two different terms is that human females have a tremendous minimum necessary investment in each of their offspring, and thus evolutionary theory predicts a much higher level of discrimination among potential mates by human females than by human males. Although men prefer women with certain attributes, and strongly prefer some attributes over others, the small minimum investment of ancestral males is predicted to have led to selection for at least some interest in mating with almost any female. Malinowski (1929, p. 292) noted that women said by Trobriand Islanders to be so physically repulsive as to be "absolutely debarred from sexual intercourse" had given birth to several offspring. Men have been known to copulate with inflatable dolls and with female calves, camels, and sheep. According to Kinsey et al. (1948), about 20 percent of men reared in rural settings admitted to a sexual encounter with a farm animal. In contrast, selection should have favored female mate choice, in which a female turns down most mating opportunities in order to pick the available male who offers the most benefits in return. This is why, as Symons (1979, p. 253) observed, among humans, "everywhere sex is understood to be something females have that males want."

Male Preferences

Since human evolutionary history involved greater competition among males for females than vice versa, it is not surprising that men exhibit a much greater desire for sex-partner variety than women (Symons 1979; Buss 1994; Townsend 1998). The male evolutionary ancestors of humans—the males who outreproduced other males in human evolutionary history—were individuals who were willing and able to copulate with many females, especially young adult females at the peak of their fertility and/or reproductive potential. Men's greater eagerness to copulate and their greater interest in and satisfaction with casual sex evolved because those traits promoted high sex-partner number in evolutionary historical settings (Symons 1979; Buss and Schmidt 1993; Townsend 1998).

In view of human evolutionary history, it is also not surprising that men are often willing to expend resources simply in order to copulate. The cross-cultural prevalence of female prostitution (Burley and Symanski 1982) reveals that copulation per se is viewed as something valuable by men. It is also not surprising that pornography disproportionately involves men paying to view the bodies and the sexual behavior of young women who are not their mates (Symons 1979). Men incapable of becoming sexually stimulated by the physical features of young adult females are probably no one's evolutionary ancestors. Prostitution and pornography vividly illustrate men's evolved motivation for high partner number without paternal investment. Both are popular with men because they provide sexual variety without commitment (Symons 1979).

Age is important in the attractiveness of both sexes, but especially in that of women, and the most attractive age in women is much younger than that in men (Symons 1979; Quinsey et al. 1993; Quinsey and Lalumiére 1995; Jones 1996). For example, men find pubescent and young adult females about equally attractive, whereas women find pubescent males unattractive but adult males attractive (Quinsey et al. 1993). Age is reflected in all external bodily features (skin texture, hair, behavior, and so on). Men's psychological adaptation for preferring young adult females evolved because of the positive relationship between fertility and young adulthood in females in the human evolutionary lineage. Males who preferred pre-reproductive-age or post-reproductive-age females were obviously outreproduced by males who considered females beautiful to the extent that their features were correlated with high fertility and reproductive potential. The evolutionary psychologist David Buss (1985, 1987, 1989) observed, in 37 samples drawn from 33 countries on six continents and five islands, a significant sex difference in attitude about the physical attractiveness (including its youth component) of a mate. That men value youth and other factors of attractiveness in mates more than women do is also documented in the cross-cultural record of traditional anthropology (Ford and Beach 1951; Symons 1979).

The sexual behavior of homosexual men—which relative to that of heterosexual men shows a much higher frequency of casual, non-committal sex and a higher number of partners—also illustrates men's evolved motivation for sexual variety without commitment (Symons 1979). Although heterosexual and homosexual men desire new sexual partners in equal number, homosexual men actually have far more new partners because their sex partners are men, who share their desire for new partners (Bailey et al. 1994).

Whereas more male reproductive effort is directed to mating effort, more female reproductive effort is allocated to parental effort. It does not follow from this that human males are incapable of parental effort. Human males, in general, are far more parental, on average, than males of the majority of mammalian species, and human males have the capacity to be as engaged in parental care as females (Geary 1998). The parental psychology and behavior of human males evolved because it promoted offspring survival and thus improved offsprings' chances of successful reproduction in the environments of human evolutionary history (Hewlett 1992). That the disparity in parental effort of the sexes controls mate choice is illustrated by men's flexibility in mate choice. When men plan to marry or otherwise engage in a long-term relationship, they are as discriminating as women in terms of such mate-choice criteria as intelligence and cooperativeness; however, when making no such investment, or when making only a limited investment, men are quite indiscriminate about a mate's intelligence or personality (Kenrick 1989).

The selection for male parental behavior in human evolutionary history had two major consequences: it created the potential for partially congruent reproductive interests of a man and a woman and it caused selection to favor different forms of sexual jealousy in men and women.

Sexual Jealousy

The evolutionary reasons for sexual jealousy are crucial to an understanding of many current social problems. Research on neglect, abuse, and murder of children shows that a major proximate cause affecting humans' decisions to expend parental effort on their putative offspring is genetic relatedness to them (Daly and Wilson 1988, 1995). Because human fertilization is internal and because pregnancy is restricted to women, men have less certainty than women of being genetically related to their putative offspring. Whereas errors in identifying maternity are almost nonexistent, a man runs a considerable chance of unknowingly treating offspring sired by another man as his own genetic offspring. (We will call this *cuckolding*, although the dictionary definition of a cuckold is merely "a man whose wife is unfaithful." The term we will use for a man's confidence that he is the genetic father of a mate's offspring is *paternity reliability*.) Cuckolding is favored by selection in a biparental species, since the cuckolder saves himself parental effort by exploiting the paternal effort of another male. In contrast, being cuckolded is selected against: not only may the cuckold fail to produce any offspring, he also expends effort on non-relatives.

The male psyche is designed, in part, to increase the probability that a man will direct parental benefits toward his genetic offspring rather than toward another male's. Thus, human anti-cuckoldry mechanisms include the emotions and behaviors associated with sexual jealousy, related forms of mate guarding, and a strong preference for fidelity in mates. In its milder forms, sexual jealousy motivates men to be vigilant for signs of their partners' sexual interest in other men. Indeed, owing to the high costs of being cuckolded, it might be said that the male mind has been selected to be adaptively paranoid when it comes to monitoring the mate's sexual interest in other men. In its extreme forms, sexual jealousy leads men to commit violence against their mates and/or against male competitors they perceive as paternity threats.

Whereas the jealousy experienced by men is focused on the act of potential or actual intercourse, the jealousy experienced by women seems to be focused on the risk of losing economic and material resources to a female competitor (Symons 1979; Daly et al. 1982; Buss et al. 1992; Geary et al. 1995). Women's sexual jealousy toward their mates could more accurately be called resource and commitment jealousy.

There is considerable evidence that sexual jealousy is more often expressed and more likely to lead to violence in men than in women (Daly and Wilson 1988). Male sexual jealousy is a major factor in wife beating and homicide in all human societies for which data exists. Males in many cultures often use violence and the threat of violence in attempts to control the sexual behavior of their mates (Flinn 1987; Smuts 1992; Wilson and Daly 1992; Jacobson and Gottman 1998). For example, in the United States, in both reported and unreported cases,[9] about 30 percent of all violence against women by single offenders is perpetrated by a husband, a former husband, a boyfriend, or a former boyfriend (Bachman and Saltzman 1995). Perpetrators in these categories kill about 28 percent of all female murder victims. Wives and girlfriends much less commonly kill their male pair-bond mates—about 3 percent of male victims (Perkins and Klaus 1996). When a woman kills her mate, it is often in the context of protecting herself from a violent, sexually jealous man (Daly and Wilson 1988; Wilson and Daly 1992). Cross-culturally, male violence toward women arises from men's psychological adaptation of sexual proprietariness, which evolved by selection in the context of paternity protection (Wilson and Daly 1992). That men use violence to control a sexual and reproductive resource that is of value to them is revealed by the fact that most of the female victims of such violence are in the age range 16–19 years, fewer in the age range 20–24, even fewer in the range 25–34, and very few above the age of 34 (Greenfield et al. 1998).

Men also attempt to control the sexual behavior of their mates by means of the economic and parental resources they provide to mates and offspring. When men have socioeconomic resources, they use them, in part, to control their mates' sexual exclusivity. Sexual proprietariness in the form of violence toward the female is strongly related to socioeconomic level, with higher rates in the lower than in the middle socioeconomic level and higher rates in the middle level than in the upper (Perkins and Klaus 1996). Women have evolved to value parental and economic resources provided by potential mates because these resources have considerable impact on a woman's ability to produce and successfully raise offspring. Men behave as if paternity reliability is something they expect in return for their commitment to transfer economic resources to a mate. Thus, it is not surprising that female sexual infidelity is a major cause of divorce in the United States (Symons 1979)—and, according to an exhaustive review by the evolutionary anthropologist Laura Betzig (1989), in all other human societies for which data exist.

Sperm Competition

The mating machinery of males illustrates the importance of adaptive design in terms of increasing paternity certainty. *Sperm competition*, a form of sexual selection, is the competition between ejaculates of different males for exclusive access to the egg(s) of a particular female (Parker 1970). This form of competition has led to a vast array of behavioral, morphological, and physiological adaptations in various species (ibid.; Thornhill and Alcock 1983; Wilson and Daly 1992; Baker and Bellis 1995; Birkhead and Møller 1992, 1998). For example, the male black-winged damselfly's penis "acts as a scrub brush" that removes nearly all competing sperm (Alcock 1993, p. 421).

That sperm competition has influenced selection in human evolutionary history is evident from the human male's ability to unconsciously adjust the size of his ejaculate depending on the threat of insemination of his mate by a sexual competitor. Because ejaculate size is highly related to how much time a pair-bonded man and woman have spent apart since their last copulation but only weakly related to cumulative time since their last copulation, men appear to use information about how much time they have been physically separated from a pair-bond mate as a surrogate for the probability of the mate's insemination by another male. The large size of the human penis and testicles relative to those of many other primates also appears to be an adaptation for sperm competition.[10]

Female Choice

Evolutionary theory predicts a greater level of mate discrimination by human females than by human males, with the expectation that female choice will often revolve around resources. The number of offspring a particular female produces is limited by environmental materials that can be converted into parental effort. Therefore, when males can effectively control these reproductive resources, females are expected to choose mates partly on the basis of these resources, preferring males with the most or the best (Bradbury and Vehrencamp 1977; Emlen and Oring 1977; Borgia 1979; Thornhill 1979). This has been shown to be the case in all species with the appropriate social systems that have been investigated to date (Thornhill and Alcock 1983; Rubenstein and Wrangham 1986).

In studies conducted by David Buss (1985, 1987, 1989), women from all over the world were found to use wealth, status, and earning potential as major criteria in mate preference, and to value those attributes in mates more than men did.[11] Wiederman and Allgeier (1992) and Townsend (1998) found that this preference not only fails to disappear among economically self-dependent women; it increases. Hence, this preference is not a product of economic dependence on males, as feminist theory might suggest.

Another variable that affects human females' choice of mates is a male's relative social status, which is a function of the degree to which he is respected by—and hence can exercise influence over—other males. High-status males have greater ability to prevent sexual encounters, including rape, between their mates and other men. The ability of headman in hunter-gatherer societies to attract mates, even though headmen seldom accumulate more resources than other men, may be attributable to the fact that their political influence enhances their ability to keep other males from raping their mates (Mesnick 1997; Wilson and Mesnick 1997).

As the evolutionary anthropologist Barbara Smuts has documented, in many human societies women are especially vulnerable to sexual coercion by men when they lack the protection of a partner (Smuts 1992; Smuts and Smuts 1993). A large study conducted by the evolutionary psychologist Margo Wilson and the ethologist Sarah Mesnick (1997) has shown that married women of all ages are less likely to be raped or otherwise sexually coerced than same-age unmarried women.

A preference for symmetric men may be related to the male's ability to protect a woman from rape. Studies by Gangestad and Thornhill (1997a) show that symmetric men are viewed by their mates as more physically protective, a factor affecting male attractiveness to women.

According to Wilson et al. (1997, p. 443), "perhaps the most important priority for many female animals in their heterosexual interactions is the maintenance of [mate] choice." According to Symons (1979, p. 92), this is the case because "throughout evolutionary history, perhaps nothing was more critical to a female's reproductive success than the circumstances surrounding copulation and conception." "A woman's reproductive success," Symons continues, "is jeopardized by anything that interferes with her ability: to conceive no children that cannot be raised; to choose the best available father for her children; to induce males to aid her and her children; to maximize the return on sexual favors she bestows and to minimize the risk of violence or withdrawal of support by her husband and kinsmen."

That women actually choose mates on the basis of resource control and status is evident from cross-cultural data from the many polygynous societies in which there is a positive relationship between the number of wives a man has and his resources and status (Betzig 1986).

A woman's attractiveness and youthfulness are major criteria affecting her ability to obtain a desirable mate in competition with other women. In monogamous societies, a woman's age and physical appearance are linked to her husband's occupational status, regardless of social class (Buss 1987; Ellis 1991; Grammer 1993; Barber 1995; Kenrick et al. 1996). The studies just cited also reveal that a woman's attractiveness is more strongly correlated with her husband's status than is her class origin or her IQ. Therefore, attractiveness appears to be an important path of upward social mobility for females, and males with higher occupational status seem capable of obtaining attractive wives. In contrast, men's physical attractiveness is not positively correlated with the status of their wives (Jones 1996).

Ultimately, the stronger sexual selection on males than on females in human evolutionary history is the reason that men strive more intensely for status and resources than women do (Alexander 1979; Browne 1995; Geary 1998). Because females prefer males with status and resources, males who had either or both during human evolutionary history had relatively high numbers of offspring as a result of both having more mates and having more mates of higher reproductive capacity (i.e., higher attractiveness, which in women is strongly related to youth and thus to fertility).

Genetic Quality

As their attraction to symmetry and other markers of health in men suggests, women may also choose partners on the basis of genetic quality as reflected by physical appearance (Benshoof and Thornhill 1979; Gangestad 1993; Thornhill and Gangestad 1993).

There are three prominent theories that attempt to explain the evolution of physical attractiveness and preferences for physically attractive mates[12]:

- According to one theory (often referred to as "good genes sexual selection"), women prefer physically attractive mates because attractive features connote genes that will contribute to the production of offspring with increased survival.
- A second theory is that women prefer physically attractive mates because their choosing such mates gives their children genes that will make them sexually attractive. This theory is usually discussed in terms of females choosing sexually attractive males so as to produce "sexy" or "sexually successful" sons. Here, "good" genes result in relatively more grandchildren for the choosy females, because their sons are likely to have more partners than the sons of females who ignore male attractiveness. According to this "sexy sons" theory, the sons' attractiveness per se, not their improved survival chances, promotes the evolutionary success of a woman's preference for attractive males.
- The third theory is that more physically attractive mates are preferred because they provide their mates and/or their offspring with more material benefits (food, protection, better territories, freedom from contagions, and so on) than less attractive mates.

All three theories have received considerable support from studies of a wide range of animal species.[13]

Recent work revealing that physical attractiveness reflects both developmental and hormonal health (Johnston and Franklin 1993; Singh 1993;

Thornhill and Møller 1997; Thornhill and Grammer 1999) supports "good genes" sexual selection in human evolutionary history.

The three most important categories of physical-attractiveness traits in human males are age, bilateral symmetry, and hormone markers. (Hormone markers are the traits of the face and other areas of the body that are proximately caused by the action of sex hormones—androgens in men, estrogens in women—during young adulthood.) All else being equal, elderly men are not preferred by women because they are less able to provide resources (including protection) and probably because their DNA has a higher incidence of deleterious mutations than that of young men (Ellegren and Fridolfsson 1997). However, older men who have accrued disproportionate wealth can be very attractive to women (Kenrick et al. 1996).

Symmetry

Bilateral symmetry is an important criterion of physical beauty across species. Some 65 studies, involving 42 species, demonstrate the importance of body symmetry in attractiveness to the opposite sex and in individual mating success (Møller and Thornhill 1998a). Indeed, symmetry may be the best available marker of the genetic and phenotypic quality of an individual organism (Gangestad and Thornhill 1999).

Asymmetry in normally bilaterally symmetric features (such as fingers and ankles) is proximately caused by the action, during development, of environmental insults, such as environmental toxins, disease, and low food quantity or quality. Genetic insults such as deleterious mutations also upset the development of perfect symmetry. That asymmetry is generated by environmental and genetic insults is well established in biology.[14] Thus, an individual's symmetry is a record of how well the individual has dealt with these insults during development and is a marker of his or her *mutational load* (the number of detrimental mutations the individual contains).[15] In various animal species, symmetric individuals survive longer, grow faster, have fewer diseases, and have more offspring than asymmetric individuals (Møller and Swaddle 1997). In humans, asymmetry is known to be associated with certain infectious diseases, with retarded development, and with reduced emotional, psychological, and physical health (Thornhill and Møller 1997; Waynforth 1998). Although asymmetry in humans increases with age after adulthood is reached, the considerable variation in asymmetry among individuals of the same age makes asymmetry an age-independent marker of phenotypic and genetic quality.

Facial symmetry in both men and women positively affects their ratings of facial attractiveness by members of the opposite sex (Grammer and Thornhill 1994; Mealey et al. 1999). In addition, according to numerous studies,[16] men's body symmetry, measured as a composite of symmetries in non-facial body parts (e.g., ears, ankles, feet, fingers, elbows), is related positively to their mating success. Relatively symmetric men self-report more sexual partners, earlier age of first intercourse, shorter time to sexual intercourse in romantic relationships,[17] and more copulations with women other than their main romantic partner (including women who are in long-term relationships with other men). In women, however, there is no correlation between body symmetry and partner number, infidelity, age of first copulation, or time to intercourse in romantic relationships.

In studies in which romantically involved couples were asked for anonymous, private reports, Thornhill et al. (1995) found that symmetric men stimulated more copulatory orgasms than asymmetric men did. Female orgasm may be a female choice adaptation, in that it appears to increase the number of sperm a woman retains after her mate's vaginal ejaculation (Baker and Bellis 1995) and to increase pair bonding by way of associated oxytocin release (Thornhill et al. 1995). In multiple-mate situations, women's copulatory orgasms therefore may favor the sperm of symmetric men over the sperm of asymmetric men and may result in preferential bonding with symmetric men (ibid.).

Because women are more discriminating about mates, a woman's number of partners depends less on sexual opportunity than a man's does. Thus, symmetry in men, but not in women, correlates significantly with all the components of mating success (extra-pair copulations, partner number, copulatory orgasm frequency, and so on).[18]

Because there is strong, consistent selection for bilateral symmetry in animals with forward locomotion (Møller and Swaddle 1997), the heritability of symmetry per se is probably near zero, since selection exhausts genetic variation in the trait under selection. The real heritability involved must be the result of genetic variation of at least two kinds: deleterious mutations that disrupt symmetrical development and genetic variation in host animals created by antagonistic coevolutionary[19] races between hosts and disease organisms. The latter is notorious for giving rise to and maintaining genetic variation in populations (Ridley 1993). Both kinds of genetic variation probably explain why health is so highly heritable. Thus, there is much opportunity for viability-based "good genes" sexual selection in humans now, and there was continuous opportunity for it throughout human evolutionary history.

Two separate studies have also found that the smell of symmetric men[20] is attractive to women not using contraceptive pills—especially near midcycle, the point of highest fertility, when choice of sire is most critical (Gangestad and Thornhill 1998; Thornhill and Gangestad 1999). That is, women find the scent of symmetric men maximally attractive when they are most likely to conceive. Women's preference for the scent of symmetric men may be an adaptation to obtain high-viability genes for their offspring. Men's scent preferences, however, are not significantly related to female symmetry. The sex difference in preference for the scent of symmetry implies that women possess adaptations that place greater importance on genetic quality of a mate, as would be anticipated from females' greater parental investment in offspring.

Additional evidence for the "good genes" theory comes from the finding that symmetric men invest less in romantic relationships than asymmetric men do (Gangestad and Thornhill 1997a). Symmetric men, who are more physically attractive to women, invest less time and money in their relationships and are less likely to be faithful and honest. Women seem to trade off time, money, and fidelity for having an attractive partner, perhaps because of the importance of having offspring with viability-enhancing genes. The only component of investment on which symmetric men score higher than asymmetric men is physical protection (as scored by both the man and the woman in the relationship). That symmetric men appear to engage in more fights with other men than asymmetric men do (Furlow et al. 1998) may be due to their greater social dominance and their larger body size (Gangestad and Thornhill 1997a).

Other Good Genes

Sex-hormone-facilitated traits are important beauty markers in humans of both sexes. At puberty and during adolescence, the ratio of estrogen to testosterone facilitates the development of sex-specific hormone markers of the face and the rest of the body. A man's relatively broad chin, long and broad lower face, broad shoulders, and increased musculature connote relatively high testosterone and thus sex-specific hormonal health. A woman's relatively short lower face, small lower jaw, and small waist connote high estrogen and thus sex-specific hormonal health and fertility.

In recent years much research has been conducted on sex-hormone traits in relation to human beauty and health.[21] Hormonal markers in humans show heritability, probably for the same reasons that body symmetry is heritable; thus, mate choice based on hormone markers may be, in part, viability-based "good genes" sexual selection.

Another aspect of human mate choice based on genetic quality affecting offspring viability pertains to mates who are genetically different from the chooser in genes at the major histocompatibility (MHC) genetic loci—that is, the sites of genes that are involved in recognition of nonself vs. self. The choice of MHC-dissimilar mates may have been driven by selection for outbreeding (i.e., avoiding mating with close genetic relatives, which leads to defective offspring). Alternatively, its function may be to increase the genetic diversity of offspring in MHC genes. Since MHC genes are fundamentally involved in recognition of infectious-disease organisms (the first stage in the host's defense), offspring of MHC-disparate parents should be better able to recognize a diversity of infectious agents. MHC-mediated mate choice appears to be based on body scent.[22]

Although human female mate preference attaches much importance to male status, male-held resources, and protection, "good genes" preferences also appear to play a significant role. In view of the big investment a female makes in the production of each baby, it is likely that female mate choice for heritable health was essential for high female reproductive success throughout human history. We have discussed female mate choice for good genes in some detail because it may have been of great importance in increasing female reproductive success in human evolutionary history. This importance identifies a fundamental problem for raped females: because it circumvents female mate choice, rape interferes with a central component of the female reproductive strategy.

Female Sexual Arousal

Female sexual arousal can be viewed as a mate-choice adaptation because a woman's sexual interest and arousal are importantly tied to a man's capability and willingness to invest in the relationship and by indications of his genetic quality (Thornhill and Furlow 1998). Women whose mate choice is circumvented, as it is in rape, rarely experience sexual arousal. Indeed, a woman's frequency of copulatory orgasm is significantly predicted by the nature of the resource environment she is in and thus by the opportunity for successful parental effort. At least in Western society, high marital happiness, substantial income, and high status of mate are associated with more copulatory orgasms in women (Fisher 1973).

As was mentioned above, female copulatory orgasm is positively correlated with the body symmetry of the mate—a pattern predicted by the hypothesis that women selectively bond with, and preferentially retain the sperm of, men of high genetic quality. Thus, degree of female sexual arousal, ranging from absence of arousal to copulatory orgasm, may be strategically related to female choice of mate and sire.

Unequal Degrees of Competition for Mates

Although selection has favored mate-preference adaptations and traits that signal reproductive potential to the opposite sex in both sexes, the degree of sexual selection is not equal. Women compete sexually (Buss 1994; Campbell 1995) but do not compete for copulation per se.

The ultimate reason why sexual competition is more intense among males than among females is that winning has influenced male reproductive success much more strongly throughout human evolutionary history. A male who prevailed in sexual competition because of his looks, status, or resource holdings was more likely to acquire multiple wives of high reproductive potential and to have had sexual access to other females. Females who prevailed in sexual competition may have boosted their reproduction—but by only a slight degree, since pregnancy is a long and energetically expensive endeavor. In other words, the range in number of potential offspring produced is much greater for men than for women. Given this circumstance, throughout human history, male humans have evolved to be more likely than females to engage in risky sexual competition that holds the potential to increase their number of partners.

3

Why Do Men Rape?

Selection favored different traits in females and males, especially when the traits were directly related to mating. Although some of these differences could have arisen from what Darwin called natural selection, most of them are now believed to have evolved through sexual selection.

The males of most species—including humans—are usually more eager to mate than the females, and this enables females to choose among males who are competing with one another for access to them. But getting chosen is not the only way to gain sexual access to females. In rape, the male circumvents the female's choice.

To appreciate the significance of female choice in human evolution, it helps to remember that adaptations evolved because they helped individuals overcome obstacles to individual reproductive success. In ancestral populations of many species, including humans, the difficulty of obtaining the parental investment of a choosy member of the other sex was a prominent obstacle to reproductive success for individuals of the sex with the lesser parental investment. That is, the difficulty of gaining sexual access to choosy females was a major obstacle to reproductive success for males. Owing to the significance of this obstacle throughout evolutionary history, there would have been strong and effective selection pressures favoring traits in males that increased their access to mates.

One means of gaining access to a selective female is to have traits that females prefer. If possession of certain resources increased a male's chances of being chosen as a sexual partner by a female, there would have been selection for males who were motivated and able to accumulate those resources. If the ability to influence other males increased a male's chances of being chosen as a sexual partner by a female, there would have been selection for males who were motivated and able to attain influential status. If success in physical competition with other males affected the number of sexual partners a male could secure, there would have been selection for traits in males that made them more successful in such competition. Perhaps most important, there would have been selection for intense sexual desires in males that motivated them to seek sexual sensations and, hence, drove them to strive in the activities that led to those sensations. And these desires would have been designed to peak in adolescence and early adulthood, when males attempt to enter the breeding population and when competition for mates is most intense.

Many traits of human males clearly are adaptations designed by sexual selection for success in obtaining resources and status and in winning various forms of male-male competition (Alexander 1979; Symons 1979; Ellis 1992; Grammer 1993; Buss 1994; Barber 1995; Betzig 1995, 1997; Geary 1998). There is also evidence that, in human evolutionary history, sexual selection favored males who exaggerated their status and their resource holdings in order to be chosen by females (Buss 1994; Geary 1998). The evidence of powerful sexual desires in males, peaking around adolescence and early adulthood, is even more overwhelming (Symons 1979; Alexander 1979).

Rape as a Type of Sexual Selection

Smuts and Smuts (1993) have suggested that sexual coercion is best conceptualized as a third type of sexual selection (in addition to mate choice and intrasex competition) rather than as merely a form of intrasex selection. (*Sexual coercion*, a broader term than *rape*, is defined as obtaining sexual access by intimidation, harassment, and/or physical force.) Like intrasex competition and intersex mate choice, sexual coercion affects differential access to mates. Of course sexual coercion interacts with the other two forms of sexual selection, but it is conceptually distinct from them for the following reason: A sexually coercive male may succeed in the competition for mates by coercing mating even though he loses in male-male competition for females and is not chosen as a mate by a female.

Because all three forms of sexual coercion—physically forced mating, harassment, and intimidation[1]—have significant survival and/or reproductive costs for females, a variety of female traits evolved because they reduced those costs. Indeed, many aspects of female social behavior—including pair bonding with a male and female-female alliances across species—may be explicable as adaptations against male sexual coercion (Smuts and Smuts 1993; Mesnick 1997).

Potential Ultimate Causes of Human Rape

As the biologist Theodosius Dobzhansky said in 1973, "nothing in biology makes sense except in the light of evolution."[2] Evolutionary theory applies to rape, as it does to other areas of human affairs, on both logical and evidentiary grounds. There is no legitimate scientific reason not to apply evolutionary or ultimate hypotheses to rape. The only scientific question concerns how to apply theoretical biology to a particular aspect of human endeavors. Evolutionary history would be applicable to human rape even if it were explicable only as a trait that exists as a result of evolutionarily novel circumstances faced by individual humans. And if such were the case, one would still want to know why men's sexual psychological adaptations are designed in a way that yields rape behavior in the novel circumstances.

As we have emphasized, in biology the term *ultimate cause* pertains to why a feature of life, including all its proximate causes, exists in the first place. Ultimate causes are sometimes broken down into *phylogenetic holdovers* and *evolutionary agents.*

Rape as Phylogenetic Holdover

The "holdover view" of rape is that men rape because males of their primate ancestors raped, the earlier primates raped because their even earlier male mammal ancestors raped, and so on. This is not a complete ultimate framework, because phylogeny (i.e., the coming into being of a species in the course of evolutionary descent from ancestral species—often called "phylogenetic inertia") is not itself an evolutionary cause of maintenance of a trait in a phylogenetically related group of organisms, such as the primates. The phylogenetic approach merely describes a pattern of evolutionary continuance of a trait; it doesn't identify the ultimate cause of the continuance.

Examining the behaviors of closely related species can give us important insights into the state of behavior in the common ancestral species of extant species. For example, the male violence seen in the great apes (humans, chimps, gorillas, orangutans) indicates that males of the ancestral species of these apes were violent (Wrangham and Peterson 1996). But an explanation of the continuance of a trait such as violence or rape requires an explanation in terms of selection of why the trait was conserved in the evolutionary history of a species.

Similarly, one cannot explicate the crossing of the digestive and respiratory tracts in vertebrates (discussed in chapter 1) in terms of ultimate cause by referring to it as a phylogenetic holdover. In this case, the ultimate causation was relentless and continual selection for respiratory and digestive function throughout the evolutionary histories of all the vertebrates. All evolutionary constraints and phylogenetic legacies ultimately involve selection in some way.

The holdover view is the basis for the popular but erroneous notion that the behaviors of non-human primates necessarily provide salient information about human psychological and behavioral adaptations. People commonly make the inference that if apes exhibit behavior X then it is part of human nature (i.e., part of the evolved human psychological architecture) and is not caused by culture. But a human psychological adaptation such as that responsible for rape must be studied in humans, and a chimp or orangutan psychological adaptation must be studied in chimps or orangutans (Thornhill 1997a). The widespread notion that studying the behavior of monkeys and apes is the best way to identify human psychological adaptation seems to arise from the erroneous dichotomy by which culture is seen as primarily or entirely characteristic of humans, and not of other primates, whereas nature is seen as biology, genes, instincts, behaviors (such as fighting) that non-human animals engage in, predispositions, and behaviors of non-human primates that seem similar to human behaviors. It is then claimed that the behaviors of non-human primates reveal human nature because culture is not a part or is only a minor part of the environment of those primates.

Rape and Evolutionary Agents

There are four evolutionary causes of trait change or trait maintenance in phylogenetic lines: selection, drift, gene flow, and mutation. In theoretical biology, the interactions and the relative potencies of these four evolutionary causes are well understood.[3]

Selection, owing to its non-random character, consistently favors traits that provide better solutions to environmental problems. Selection is so powerful that modern biologists realize that using the other evolutionary causes to explain anything about life requires one to reconcile the non-selection explanation with the overwhelming power of selection. This is why the biologist Graham Bell (1997) calls selection "the mechanism of evolution." With this in mind, let us consider the various alternative ultimate hypotheses regarding human rape, many of which can be rejected on the basis of existing evidence (Palmer 1988b; Thornhill 1999).

One possibility is *mutation-selection balance*, in which a maladaptive trait continually arises by means of a very low rate of mutation and is selected against as it appears. But mutation-selection balance can account only for very rare traits—traits found in less than 1 percent of the population.[4] Although a particular rape might be related to a specific recent mutation, rape is far too common to be explained by mutation-selection balance. First, rape appears to occur in all human societies. Cross-culturally, it is not uncommon in either modern or pre-industrial societies.[5] During war, it is quite common (Brownmiller 1975; Morris 1996; Cheng 1997; Littlewood 1997). Also, women's apparent adaptation to deal with rape (discussed in chapter 4 below) implies that rape has been common enough in human evolutionary history to select for counter-adaptations in women. Thus, mutation-selection balance cannot explain human rape. Rape appears to have been a consistent part of the social environment throughout human evolutionary history.

A second possibility is *drift* (chance variation in reproduction among individuals). But, as we mentioned in chapter 1, only traits that lack significant costs to reproductive success can be attributed to drift. Only selection explains why traits that are costly and that occur at greater than 1 percent frequency exist despite their costs. Rape has major costs to the rapist: time, energy, and (especially) risks of physical injury and social ostracism. Thus, drift apparently cannot account for rape's existence.

Third, perhaps human rape is the result of an evolutionarily novel environment. For example, one might imagine that men's consensual sexual adaptations are not adapted to some feature of the modern environment (e.g., plastics, pollution, abnormally high population densities), and that this somehow yields rape. Roger Masters and his colleagues, who have amassed data indicating that levels of heavy metals may predict crime rates across US counties, propose that heavy metals—especially lead—disrupt psychological adaptations of impulse control and thereby lead to a greater rate of criminality (Masters et al. 1999). Lead may account for certain cases of rape, just as mutation may. However, as a general explanation for why men rape, the lead hypothesis is not compatible with the apparent existence of rape in all cultures, including hunter-gatherer societies. Rape's universality indicates reliably that it arises from a wide range of developmental environments and that it is not tied to society-specific evolutionary novelties. Also, the evidence of female adaptation against it reveals that rape is not a new event generated by new circumstances in the human environment. And the fact that rape is seen in many non-human species is further evidence that evolutionary novelty is not a useful general explanation.

In the older literature on human rape (e.g., Rada 1978b), it was repeatedly proposed that rape is caused by some type of unusual pathology. This view is subsumed by our discussions of mutation-selection balance and novel environments.

Another ultimate possibility is that "rape may be an evolved male mechanism whose primary aim is not fertilization in the present, but *control*—for the ultimate purpose of fertilization in the future" (Wrangham and Peterson 1996, p. 141). Suggested as an explanation of rape in non-human primates (Rijksen 1978; Smuts and Smuts 1993) and in birds (Gowaty and Buschhaus 1998), this theory proposes that the establishment of dominance through demonstration of the male's ability to physically control the female could eventually lead to greater reproductive success for some or all males through various effects on female behavior. For example, it could force females to trade sexual access to otherwise unchosen males for protection from raping males (Gowaty and Buschhaus 1998). Or it could cause females to acquiescence to future mating attempts by undesired males because the female has learned she cannot prevent a rape and hence has nothing to gain by resisting.

The preceding explanation has a serious theoretical weakness that may make it inapplicable to any species: If demonstration of the male's ability to physically control the female has the proposed effect on the female's behavior, why do the males commit rape instead of just non-sexual aggression? This explanation, when applied to birds, really isn't an explanation of rape at all; it is only an explanation of male aggression, because it is the demonstration of aggression that is claimed to influence the behavior of females in a way that leads to future sexual access. Gowaty and Buschhaus (1998, p. 218) put it as follows: "This idea suggests that it is possible that some observations of aggressive 'copulations' were just as likely to be male aggression against females rather than copulation attempts."[6]

Of course, one could argue that the behavior of females might be more influenced by rape than by non-sexual assault because females find being raped more unpleasant than simply being beaten. Indeed, we believe that rape is a more traumatic experience, and we maintain that this is because throughout evolution being raped led to the additional negative fitness consequences of being fertilized by undesired males. The "dominance explanation" actually implies the alternative "immediate-fertilization" explanation (discussed below). Indeed, Gowaty and Buschhaus's main basis for proposing the dominance explanation of rape in birds is the existence of complexly designed "counter-mechanisms" in females to prevent fertilization through forced copulation. What Gowaty and Buschhaus fail to point out is that, if these mechanisms were designed to counter insemination via forced copulation, their existence implies that the possibility of females' being immediately fertilized by raping males was a significant selective pressure in the evolutionary history of these birds.

The dominance explanation of rape is based on evidence of one possible *effect* of rape in some species (a change in the behavior of females), but it is not supported by any evidence of rape's being functionally *designed* to have such an effect.[7]

Human Rape: Adaptation or By-Product?

There are currently only two likely candidates for ultimate causes of human rape:

- It may be an adaptation that was directly favored by selection because it increased male reproductive success by way of increasing mate number. That is, there may be psychological mechanisms designed specifically to influence males to rape in ways that would have produced a net reproductive benefit in the past. "How could rape increase reproductive success?" ask Wrangham and Peterson (1996, p. 138). "There is," they continue, "a blindingly obvious and direct possibility: By raping, the rapist may fertilize the female." Remember, however, that identifying an *effect* that may have increased reproductive success in past environments is not the same as identifying the *function* of an adaptation.

- It may be only a by-product of other psychological adaptations, especially those that function to produce the sexual desires of males for multiple partners without commitment. In this case, there would not be any psychological mechanism designed specifically to influence males to rape in ways that would have produced a net reproductive benefit in the past.

There are reasons for seriously considering each of these hypotheses. On one hand, rape is usually costly in evolutionary terms, owing primarily to potential punishment of the rapist or to his potential injury by the victim or by her social allies. When associated with any trait, such costs imply that the trait has had overriding reproductive benefits. The existence of such costs might be expected to act as a selective pressure, producing psychological mechanisms that caused males to be more likely to rape when the potential costs were low. On the other hand, many human behaviors other than rape clearly are by-products of the intense sexual desires of human males and the sexual choosiness of human females: Sexual abuse of children can be seen as an example of males attempting to gain sexual access to individuals who, because of their age, are relatively unable to control sexual access. Bestiality is a means of experiencing sexual stimulation somewhat like that experienced in intercourse with a human female without having to be chosen by one. Frottage (rubbing a woman's body through her clothing, usually in crowded quarters such as an elevator) and genital exhibitionism give sexual stimulation to male perpetrators by circumventing female choice. Masturbation—far more common among males than among females—is the most widespread male behavior that can be seen as a means of obtaining sexual stimulation without being chosen by a human female as a sexual partner. Although all these acts are examples of males' attempting to gain sexual gratification without meeting the criteria of an adult human female's mate choice, none of them are likely to be adaptations. They are apparently all merely by-products of the adaptations governing male sexual desires.[8]

We will now examine the evidence concerning whether human rape is a product of adaptations designed specifically to increase a male's reproductive success or whether it is a by-product of adaptations designed for attaining sexual access to consenting partners. Although we two authors disagree as to which of these two ultimate explanations of rape we expect to be confirmed by evidence (Thornhill 1980; Palmer 1988b, 1989a, 1991, 1992b; Thornhill and Thornhill 1983, 1992a,b), we are now in agreement with the statement that "whether there exist psychological adaptations specifically for sexual coercion [or rape], adaptations that entail something more than the simultaneous arousal of sexual and coercive inclinations, has yet to be elucidated" (Wilson et al. 1997, p. 453). Our goals here will be to describe the best candidates for rape adaptations and to describe the evidence that may be garnered in the future to settle the question. In keeping with Williams's (1966, 1992) view that complex traits should be considered adaptations only if they cannot be accounted for as by-products, we will examine the by-product hypothesis first.

The by-product hypothesis was first suggested by Symons (1979, p. 284): "I do not believe that available data are even close to sufficient to warrant the conclusion that rape itself is a facultative adaptation in the human male. . . ." Remember that by-product explanations are still evolutionary and still require the identification of the adaptations proposed to have generated the trait in question as a by-product (Thornhill et al. 1986; Palmer 1991; Thornhill and Thornhill 1992a,b). Identification of connections to these underlying adaptations requires identification of the proximate mechanisms involved. This is why Losco (1981, p. 336) stated that "insufficient attention to proximate concerns may result in a misreading of the forces at work in the selection process" and that "researchers who concentrate exclusively on ultimate function run the risk of cataloguing behaviors as 'adapted' which may instead be by-products or secondary effects of other adaptations."

The features that Symons proposed as underlying rape were the evolved male-female differences in *sexual* desire. Several later evolutionary models of rape suggested that rape is motivated both by sexual desire and by a "drive to possess and control" (Ellis 1989, 1991).[9] Certainly, people in general—men, women, and children—strive to possess and control numerous aspects of their lives (Geary 1998), and a drive to possess and control a mate's sexuality is seen in mate-guarding behaviors of human males, including claustration (physically isolating a female from men other than her husband and her close male relatives), clitoridectomy, and sexual jealousy (Wilson and Daly 1992; Buss 1994). These drives, however, appear to be associated with rape only rarely. Except in some cases of marital rape (discussed below), rapists do not typically seclude and jealously guard their victims for prolonged periods of time. Hence, the question of sexual motivation has crucial theoretical consequences as well as practical consequences. Specifically, Symons proposed that the primary adaptations responsible for the occurrence of rape were the mechanisms involved in the human male's greater visual sexual arousal, greater autonomous sex drive, reduced ability to abstain from sexual activity, much greater desire for sexual variety per se, greater willingness to engage in impersonal sex, and less discriminating criteria for sexual partners (Symons 1979, pp. 264–267). Symons (ibid., p. 267) wrote that, as a result of these sexually selected adaptations, designed to increase men's mating success by increasing the number of sexual partners that men acquire, "the typical male is at least slightly sexually attracted to most females, whereas the typical female is not sexually attracted to most males." Symons felt that rape is a by-product, or a side effect, of the adaptations producing this situation—that is, that rape is not itself an adaptation, because none of the evolved mechanisms involved in it were selected specifically for rape. Instead, the mechanisms exist because of their promotion of male reproductive success in contexts other than rape.

To support the alternative view requires identification of mechanisms involved in rape that were designed by selection in the past specifically for reproduction by means of rape. Keep in mind that mechanisms merely allowing (or even increasing) a male's ability to rape are not necessarily designed for that purpose. For example, the larger average size of human males certainly makes it easier for them to commit rape; however, it cannot be considered an adaptation designed by sexual selection to promote rape, because such sexual dimorphism is much more parsimoniously explained as an adaptation for intramale competition in polygynous species (Alexander et al. 1979). Similarly, "the very fact that men are able to maintain sexual arousal and copulate with unwilling women requires an explanation, for such persistence without cooperation or encouragement is evidently not a universal feature of male sexual psychology in all animal species" (Wilson et al. 1997, pp. 457–458). It does not necessarily follow, however, that this must be attributable to the existence of psychological mechanisms *designed* for such copulations. Malamuth (1996, p. 276) suggests that it is "because of their greater capacity for impersonal sex" that "men can be fully sexually functional in the face of an unwilling sexual partner who has no emotional desire for or bonds with the male." That is, the ability to copulate with unwilling females may be a by-product of other adaptations—including those of men's great interest in and pursuit of impersonal sex—that exist because of their promotion of male reproductive success in contexts other than rape.

What is needed to support the human rape-adaptation hypothesis is evidence of a phenotypic feature in the human male analogous to the notal organ possessed by the males of certain scorpionflies.[10] This organ—a clamp located on the top of the abdomen, behind the wings—appears to have been designed specifically for rape. Not only does it accomplish the task in a precise, efficient, and complex manner; it does not appear to be used in any other activity. Is there evidence of any analogous mechanism in the human male? Such evidence would include some aspect of rape that cannot simply be accounted for by the differences in male and female sexuality described by Symons. Before we examine some candidates for men's rape adaptations, we will discuss some of the best-known rape adaptations in animals in order to clarify what a rape adaptation is.

A male scorpionfly obtains mating either by offering the female a nuptial gift (a mass of hardened saliva that he has produced, or a dead insect) or by force (Thornhill 1979, 1980, 1981, 1984, 1987). Females prefer as mates males bearing nuptial gifts; they flee when approached by a male not bearing a gift, but they approach a male who has one. A male without a nuptial gift approaches a female, grabs her with his genital claspers (a pair of clamp-like structures, one on either side of the penis), and positions the anterior edge of one of her forewings in his notal organ, where he then holds it throughout mating. Approaching a female and grasping her with the genital claspers show no functional design for rape. The behavior of approaching another individual is not specifically for rape, and males even use the genital claspers in fights with other males. However, the notal organ is used specifically for rape. In an experiment, males whose notal organs were rendered inoperational by the application of beeswax were unable to obtain forced mating, but control males with operational notal organs were successful in forced copulations; moreover, when males with non-functional notal organs were allowed to present nuptial gifts to females they mated successfully (Thornhill 1980, 1984). Other experiments showed that the notal organ is not necessary for full insemination of the female in unforced mating. Thus, the notal organ is not an adaptation for sperm transfer. Nor does the organ improve a male's chances of retaining a female during mating when another male intrudes; hence, it is not an adaptation for keeping intruding males from disrupting copulation. Additional hypotheses for the notal organ that were alternatives to the rape-adaptation hypothesis have been tested and rejected. All evidence favors the hypothesis that the notal organ is designed for rape: it functions to secure a mating with an unwilling female and to retain her in copulation for the period needed for full insemination (Thornhill and Sauer 1991).

Special-purpose male adaptations for coercive mating also occur in other insects. Arnqvist (1989), who studied a species of waterstrider using experimental techniques similar to those used in the above-mentioned experiments with scorpionflies, found that the two projections on the bottom of the male waterstrider's abdomen were specifically designed to obtain mating by force. The male sagebrush cricket also has specialized appendages that function in rape (Sakaluk et al. 1995).

As we have emphasized, adaptations are phenotypic solutions to environmental problems that affected the reproductive success of individuals during evolutionary history. Thus, to discover the functional design of an adaptation is to discover the kind of selection that shaped the adaptation. The existence of rape adaptation in the insect species discussed above reveals two things about their evolutionary history: that rape occurred consistently in the past and that it increased the mating and the reproductive success of males that raped relative to those that did not rape. That is, the male ancestors of modern scorpionflies and waterstriders had the capacity to rape adaptively.

Potential Rape Adaptations of Men

Men obviously don't have a clamp designed specifically for rape, nor do they have any other conspicuous morphology that might be a rape adaptation. We must therefore look to the male psyche for candidates for rape adaptations. If found, such adaptations would be analogous to those in the male insects.

Analogous is a technical term in evolutionary biology. Two analogous adaptations are products of the same historical selection pressure (in this case, selection in the context of physically forced mating), but the selection molds different phenotypic features to accomplish the same function. The wings of birds and those of insects are familiar examples of analogous adaptations for flight. Selection in the context of flight ability created both, but from very different anatomical substrates.

We will examine some possible human rape adaptations by assessing whether, in each case, there is evidence of special-purpose adaptation for rape. If these psychological adaptations currently exist, it is because they influenced males in ancestral populations to rape when the ultimate benefit (production of offspring) outweighed the ultimate costs (negative fitness consequences resulting from injury, punishment, etc.). However, like all adaptations, they could have accomplished their ultimate function only through proximate mechanisms. Hence, determining the actual existence of these adaptations requires us to evaluate the evidence for psychological mechanisms designed to respond to environmental variables influencing the proximate costs (e.g., painful sensations resulting from injury or punishment) and the proximate benefits (e.g., pleasurable sexual sensations) of rape.

One, or more, or none of the following potential adaptations may exist:

- psychological mechanisms that help males evaluate the vulnerability of potential rape victims
- psychological mechanisms that motivate men who lack sexual access to females (or who lack sufficient resources) to rape
- psychological mechanisms that cause males to evaluate sexual attractiveness (as indicated by age) differently for rape victims than for consensual sexual partners
- psychological and/or other physiological mechanisms that result in differences between the sperm counts of ejaculates produced during rape and those of ejaculates produced during consensual copulation
- psychological mechanisms that produce differences between the sexual arousal of males caused by depictions of rape and that caused by depictions of consensual mating

• psychological or other mechanisms that motivate males to engage in rape under conditions of sperm competition.

Evaluation of vulnerability

According to Shields and Shields (1983, p. 115), rape "is expected to occur only when its potential benefit (production of an extra offspring) exceeds its potential cost (energy expended and risk taken) owing to some probability of resistance or retribution that would reduce a rapist's (or his kin's) reproductive success." The hypothesis yields the prediction of a psychological mechanism in men functionally specialized to evaluate the vulnerability of females to rape, as opposed to a broader mechanism designed to calculate the benefits and costs of other social transgressions (e.g., theft, murder). Rape-specific costs to the rapist include injury by the victim and retribution by the victim's kin and/or by other individuals. A demonstration that men were specifically motivated to rape by conditions related to its low cost would imply a rape-specific motivational mechanism.

That men are more likely to rape when rape's proximate benefits exceed the chances of injury and punishment is evident from the cross-cultural pattern of laws and uncodified rules against rape and the associated penalties (Palmer 1989a). The detailed laws pertaining to rape in the Bible and related documents and the designation of rape as a capital offense in some nation-states (Hartung 1992 and personal communication) reveal knowledge that men are more likely to rape when costs are low. Rape in war also shows that men pay attention to benefits and costs during rape. The war context provides the proximate benefit of mating with young and thus attractive women. The proximate costs of rape are low, since there is little or no protection of the female targets by family and mates and since punishment for rape is relatively unlikely. However, the current evidence on this type of focused rape adaptation is not sufficient to support the rape-specific-adaptation hypothesis. Theft too is rampant during war, and for the same reason that rape is: benefits are gained at low cost. Similarly, severe sanctions against rape across cultures do not, in themselves, provide evidence that men possess a psychological adaptation designed to assess the costs and benefits specific to rape. That is, these patterns may be attributable to a cost-benefit-evaluation mechanism that is not specific to rape.

Males' lack of resources and/or lack of sexual access to females
Another possibility is a mechanism that lowers the threshold for rape in males who lack alternative reproductive options (Thornhill 1980; Thornhill and Thornhill 1983). A lack of alternative reproductive options correlates with the male's inability to acquire resources. Hence, a key prediction from this hypothesis—sometimes referred to as the *mate-deprivation hypothesis* (Lalumiére et al. 1996)—would be evidence of a psychological mechanism designed to make rape contingent upon lack of resources and/or lack of sexual access to females.

This hypothesis is supported by evidence that rape is disproportionately committed by males with lower socioeconomic status, insofar as that is evidenced by data on rapists in the penal system (Thornhill and Thornhill 1983). The distribution of reported and unreported rapes in relation to the household income of victims also provides support. For both rape and attempted rape, the correlation between household income and number of rape victims per 1000 people in the population aged 12 and over is negative and linear. That is, higher rape rates are seen in lower-income homes and areas, and the lower the income the higher the rate (Perkins et al. 1996). Moreover, Kalichman et al. (1998) report that 42 percent of surveyed women living in low-income housing in the state of Georgia had been raped, and Eisenhower (1969) reported that a female residing in an inner city stood a 1/77 chance of being raped whereas in the "more affluent areas" the risk was 1/2000 and in the richer suburbs it was 1/10,000. Also, women of lower socioeconomic levels are more fearful of rape (Pawson and Banks 1993). This pattern, coupled with the strong tendency (82 percent) for a rapist and his victim to live within the same vicinity (Amir 1971), indicates that men of lower socioeconomic status are overrepresented among rapists.

A study of adolescent male sexual criminals by the evolutionary psychologist A. J. Figueredo and his colleagues (Figueredo et al. 1999) supports the prediction that males of lower socioeconomic status are more likely to rape. The offenders Figueredo et al. studied included rapists and were characterized by backgrounds of repeated frustration and failed romantic and sexual relationships. They had lower psycho-social functioning, including learning disabilities and psychological disorders that may have generated the competitive disadvantage they exhibited.

There are, however, a number of types of evidence that prevent the general pattern of rapists as socially disfranchised males from establishing the predicted mechanism. First, rape is far from being an exclusive act of low-status males; there are many instances of rape by high-status males with high access to females for consensual sex. Second, the correlation between low status and crime is not limited to rape; low-status males disproportionately commit many other kinds of crimes (Alexander 1979; Daly and Wilson 1988; Weisfeld 1994).[11] Third, self-report studies of men have found a positive correlation in normal, unincarcerated men between sexually coercive tendencies and high level of sexual access to females (Lalumiére et al. 1996; Malamuth 1998).

That high-status men with sexual access to females sometimes rape refutes a simplistic version of the mate-deprivation hypothesis, but it does not rule out the possibility that a mechanism making males with little access to resources more likely to rape may be one of the psychological mechanisms influencing rape. Rape by men with high status and abundant resources may arise from a combination of impunity and the hypothetical adaptation pertaining to evaluation of a victim's vulnerability. If so, their raping must result from adaptations other than that suggested by the second hypothesis; however, the proposed adaptation might still account for the raping behavior of males who lack resources.

There is some tentative evidence as to what psychological mechanisms might be involved in the raping behavior of males with resources and sexual access to females. Although men's own reports of rape are difficult to evaluate, the various findings mentioned above suggest that a lack of sexual restraint may correlate with the likelihood of high-status males' committing rape. The evolutionary psychologist Neil Malamuth (1996, 1998) has done extensive research on reduced sexual restraint and related variables in regard to sexual coercion by men.[12] His approach focuses on how the integration of different kinds of psychological adaptations of men may cause rape, with an emphasis on psychological mechanisms underlying the rewards men derive from impersonal sex and from sexual proprietariness. They hypothesize that individual men's sexual impulsiveness and risk taking are set during development through experiences related to the environment of upbringing. The developmental experiences felt to be most important are reduced parental investment (resulting from poverty or from the absence of the father) and a rearing environment in which social relationships in general are not enduring or committed. It is well known that these experiences tend to co-occur in rearing environments and are felt to mold an individual's sexual strategy, leading to earlier sexual maturity and intercourse in both sexes and, in boys, to disenfranchisement and juvenile delinquency (MacDonald 1988, 1992; Surbey 1990; Lykken 1995; Barber 1998; Malamuth and Heilmann 1998). One developmental experience emphasized by Malamuth and his colleagues is a male's perception of rejection by potential mates and an associated history of non-committal heterosexual mateships. Men emerge from this developmental background with a perception of reduced ability to invest in women, an expectation of brief sexual relationships with women, a reduced ability to form enduring relationships, a coercive sexual attitude toward women, and an acceptance of aggression as a tactic for obtaining desired goals.

This model, then, is a version of the mate-deprivation hypothesis that emphasizes lack of enduring and committed sexual relationships rather than lack of sexual access. It is an explicit evolutionary developmental model of men's sexual coercion. Malamuth has tested some of its central predictions and has found support for it in his ability to identify men with sexually coercive tendencies. The model, however, can be viewed as a hypothesis that rape is an incidental effect of men's sexual adaptations other than rape, or as a hypothesis that rape reflects a particular developmental adaptation to stimuli associated with low probabilities of forming long-term relationships with women.

Malamuth (1996) comments that it is likely that the constellation of male adaptations he suggests as being involved in rape would have led to increased male reproductive success under some circumstances during human evolutionary history. This is probably correct. The issue we are treating here, however, is not whether rape *ever* increased male reproductive success; it is whether such effects were frequent enough and strong enough to have led to demonstrable rape-specific adaptation. It should also be noted that counter-selection in other contexts or lack of heritability can prevent a selective pressure from actually leading to adaptation to cope with the ecological problem generating the selection (Thornhill 1990).

Research such as that of the evolutionary psychologist Martin Lalumiére and his colleagues (1996) on men's self-reported tendencies toward rape is problematic in view of female mate choice. Lalumiére et al. found that men with self-reported histories of high partner number also report more rape-related behavior. One probable reason such men are likely to have had many partners is their physical attractiveness to women—an important influence on the number of women willing to have sex with a man (Gangestad and Thornhill 1997a,b, 1998). There is also evidence that men with many partners are less committed in their heterosexual relationships than other men and yet are attractive to women (Gangestad and Thornhill 1997a). As we discussed earlier, physical attractiveness in men may have connoted genetic quality pertaining to offspring survival in human evolutionary history. Although women sexually desire physically attractive men, they may receive few material benefits from them. The female's strategy might, therefore, include displaying to physically attractive males an unwillingness to mate. This display may function as a signal to the male that the female is discriminating about mates, which may increase the man's perception of her value in terms of paternity reliability, and thus may result in her eventually getting more material benefits from the male than she would get in the absence of the display. If a woman's display of reluctance is truly effective, a man who achieves copulation with her will perceive that he achieved it by force. Empirical determination of whether this is part of the pattern reported by Lalumiére and colleagues would involve testing the following predictions: The copulations that attractive men report as forced will sometimes, or typically, involve significant female sexual arousal, including orgasm. Also, women in such situations will often desire to continue dating such physically attractive men even after the seemingly coercive sex. With the cooperation of the female victims in these cases, these matters could be explored empirically.

The mixed evidence with regard to the existence of a psychological adaptation connecting reduced sexual access to females or reduced resource holding with propensity to rape calls for further research to explore this candidate for a rape adaptation.

Choosing victims

Another candidate for a rape adaptation would be a victim-preference mechanism designed during human evolutionary history specifically to maximize the reproductive benefits of rape. Because the chance that a one-time copulation will result in a viable offspring correlates with the female's fertility (as opposed to her future reproductive potential—i.e., reproductive value), selection might have produced a psychological mechanism influencing males to be more likely to rape highly fertile females. Most likely this ultimate benefit would have been accomplished by means of a proximate "beauty-detection" mechanism, designed specifically for rape, that would motivate males to prefer to rape females at the age of peak fertility (in their early to mid twenties in the present-day United States, with similar peaks in other societies, including hunter-gatherer societies) (Thornhill and Thornhill 1983; Symons 1995; Hill and Hurtado 1996). In contrast, the by-product-of-evolved-sexual-differences hypothesis would predict the same male preference with regard to rape victims as with regard to other sexual partners. Men's preference for consenting sexual partners appears to be closer to the age of peak reproductive value (the mid to late teens) than to the age of peak fertility (Symons 1979, 1995; Johnston and Franklin 1993; Quinsey et al. 1993; Jones 1996; Perrett et al. 1998).

Thornhill and Thornhill (1983) attempted an initial test of this hypothesis and found that the age of US rape victims does correlate slightly better with age of peak fertility than with age of peak reproductive potential. However, more research is required to measure the effect of differences in vulnerability of victims in the relevant age categories. Because both the proximate costs associated with vulnerability and the proximate benefits of sexual desirability importantly influence men's selection of rape victims, differences in vulnerability (e.g., frequency of being isolated with males) among women at different ages may explain why the age distribution is slightly closer to peak fertility than to peak reproductive potential (Palmer 1992a). Further, even if the greater correlation with fertility than with reproductive value is due to greater desirability, it is still possible that the mechanism involved has been designed for selecting consenting one-time sex partners or other short-term mates rather than rape victims, because a high-fertility choice would have increased male reproductive success in any short-term mating situation in human evolutionary history. There is no question, however, that rapists primarily target females of fertile ages. This pattern is seen in every available set of data on female rape victimization.

Thornhill and Thornhill (1983) tabulated all the major US data sets available at that time on female rape victim ages. They also commented on several additional but more limited data sets from other industrial societies. They concluded that young women are greatly overrepresented and that girls and older women are greatly underrepresented in the data on victims of rape. The authors cautioned that these data were based primarily on *reported* rapes and may thus have been biased. However, numerous later studies indicate that both reported and unreported rapes show the same age pattern. One national study of reported and unreported rape included a representative sample of women 18 and older and found that 62 percent of the victims (at the time of the rape) were of ages 11–29, only 6 percent were older, and 29 percent were below 11 (Kilpatrick et al. 1992). However, these data were not broken down by the nature of the rape (which was defined broadly to include any sexual penetration, by finger, object, or penis, of mouth, rectum, or vagina), nor were data collected on the proportion of the victims under 11 who were exhibiting secondary sexual traits (e.g., estrogen-facilitated development of breasts, buttocks, and/or thighs). Men's evolved sexual psychology is predicted to be more sexually motivated when the latter traits are present. Meeting this prediction, studies in which data on the ages of female rape victims under 15 are broken down by year show increased rape victimization with increased age (Hursch 1977).[13] The increasingly early age of menarche in Western females (Barber 1998) contributes to the enhanced sexual attractiveness of some females under 12. Also, the youngest rape victims are raped in proportion to their occurrence in the population: child (defined as under 12) rape victims comprised an estimated 16 percent of US rape victims in 1992, when females under 12 comprised 17 percent of the US female population (Langan and Harlow 1994).

Another large study of reported and unreported rapes and other sexual assaults in a representative sample of US females 12 and older—the National Crime Victimization Survey Report of data for 1993 (Perkins et al. 1996)—showed that population-based rates were highest in the age range 16–24 and next-highest in the range 12–15. Rates decreased in each higher age range after 24, and there were few cases in which the victim was older than 50. A similar study for 1994 (Perkins and Klaus 1996) revealed exactly the same pattern, and earlier National Crime Victimization Survey Reports show the same pattern. In analyses of data on attempted and completed rapes for the years 1973–1982, the ages of female rape victims ranged from 12 to 96, the median age was 21, and 92 percent of the victims were 40 or younger (Felson and Krohn 1990). The average age of female victims of *robbery and* rape (28) was significantly younger than the average age of female victims of robbery only (35)—that is, when the victim of a male robber was young, the robber was more likely to rape her. As Greenfield (1997) found when he reviewed more than two dozen data sets maintained by the Bureau of Justice Statistics and the Federal Bureau of Investigation, the same pattern—great overrepresentation of adolescent and young adult female victims—is seen in all available data sets involving only rapes reported to police or involving ages of victims of imprisoned sex offenders. Data on female rape victims' ages during wars (across societies and over considerable time spans) also show that most were young.[14]

Rapes and other sexual assaults of males by males constitute only about 1–3 percent of sexual assaults, but data show that these sexual assaulters also prefer youthful features in their victims (Perkins et al. 1996). This pattern is likely to be a by-product of men's evolved preference for young sex partners (Symons 1979; Quinsey et al. 1993; Quinsey and Lalumiére 1995).

We are not claiming that the available data on rape victims' ages are perfect depictions of rapists' sexual desires. (Presumably, rapists weigh benefits and costs and select victims accordingly, just as other people select from available options in sexual and non-sexual domains of life.) Nor are we claiming that the data are without bias. Rape probably remains significantly underreported, even in surveys that strive to obtain the highest degree of accuracy.[15] There is some evidence that young women, relative to post-reproductive-age women, are more likely to desire to keep a rape secret (Thornhill and Thornhill 1990a). Thus, it may be that young women's rapes are most subject to underreporting and hence even more disproportionately frequent than the studies based on reported rapes suggest. And false accusations (Kanin 1994) may bias the data on purported rape victimization if such accusations are a function of age. Despite these concerns, however, we are safe in concluding that young adult females are vastly overrepresented and that female children and post-reproductive-age females are greatly underrepresented in the population of rape victims. This pattern has been shown so many times, across so many settings, by so many methods, that it is established beyond any reasonable doubt.

Sperm counts

Recent advances in the study of sperm competition may offer a way to search for an adaptation in human males designed specifically to promote reproduction via rape. Such a mechanism would cause ejaculates produced during rape to differ from those produced during sex with consenting partners in a manner conducive to high probability of fertilization during rape. As we mentioned in chapter 2, men possess ejaculatory mechanisms that adjust sperm volume in response to how long they have been apart from pair-bond mates, a variable that affects the potential for sperm competition (Baker and Bellis 1989, 1993, 1995). Of course, it is not likely that the evidence needed to determine if sperm counts during rape differ significantly from those during encounters with consenting partners will ever be gathered. But, though it would be less conclusive, perhaps evidence on sperm counts could be gathered from studies in which subjects would masturbate to audio and video depictions of rapes and of consensual sex acts. Men are easily sexually aroused by sexually explicit visual and auditory cues, and their sexual arousal in response to a depiction of rape versus a depiction of consensual sex has been well studied in laboratory experiments (Lalumiére and Quinsey 1994; Thornhill and Thornhill 1992a,b; Harris and Rice 1996). This same general methodology could be used to determine if ejaculate size varies when men masturbate to depictions of rape and to depictions of consensual sex.

If the hypothesized adaptation exists, rapists are predicted to deliver large ejaculates because rape would consistently have been associated with high sperm competition in human evolutionary history. The woman's resistance during rape is expected to be perceived by the rapist as indicating that she has an investing consensual mate. Also, rape in warfare is expected to have often involved multiple men inseminating a victim over a short period of time, with men often copulating in the presence of other men and placing their sperm in competition with the sperm of others. If men produce larger ejaculates when viewing depictions of rape than when viewing depictions of a woman presented as engaging in consensual extra-pair copulation (and thus having a pair-bond sexual partner), ejaculate adaptation to rape would be indicated. This contrast not only compares rape with consensual copulation; it also attempts to control for evolved perception of sperm competition by men in both contexts.

Patterns of sexual arousal

Another possible adaptation would involve changes in the arousal patterns of males during rape relative to during other sex acts. One possible adaptation might have evolved in response to rape's potentially high costs—costs corresponding, in part, to the chances of being observed. The risk of detection should correlate with the duration of the act. Hence, males might have been selected for quicker penile arousal and ejaculation during rape than during consensual sex. However, selection may have favored quick ejaculation even during consensual sex, especially with women pair-bonded to other men. Indeed, the evolutionary psychiatrist Randy Nesse and the biologist George Williams (1994) have suggested that the greater frequency of "premature" ejaculation among younger males may actually be an adaptation promoting reproduction by males who are too young to be pair-bonded and whose only sexual access to females would be with women pair-bonded to other men. Hence, it would be necessary to control for age and to demonstrate that males ejaculate more quickly during rape than during other intercourse, including intercourse with married women. This possibility, like the one discussed above, could be examined by means of laboratory studies involving video and audio cues.

Another experimental approach that may be useful for discriminating between the hypothesis of incidental effect and the hypothesis of rape adaptation involves men's sexual arousal arising from the physical control of an unwilling sexual partner. According to the hypothesis of rape adaptation, gaining physical control over an unwilling sexual partner is expected to be sexually arousing to men because it facilitates rape. As long as she is capable of resisting, a potential victim might injure her assailant and thwart an assault. Rape-specific selection is anticipated to have generated mechanisms for assessing such risks and regulating a male's rape motivation and behavior accordingly. In the context of coercion, male sexual arousal has a cost: the man's sexual focus and his relative unawareness of surroundings during high sexual arousal might interfere with his ability to detect the presence of dangerous conspecifics or predators (Thornhill and Thornhill 1992a). In contrast, the side-effect hypothesis does not predict that gaining physical control of and subduing an unwilling partner will be more sexually arousing than completing a rape despite the victim's continued resistance. Some evidence that men's sexual arousal is enhanced by physical control of an unwilling woman has been obtained from laboratory experiments in which men's arousal is measured as they view video depictions of consensual sex involving physically restrained women and of rape involving physically restrained women (Thornhill and Thornhill 1992b).

This is not to say that men are sexually aroused by violence per se. They aren't (Thornhill and Thornhill 1992a; Lohr et al. 1997; Quinsey et al. 1984). Nor is it to say that the motivation to dominate and brutalize the victim is paramount, or even necessary, in rape causation. Rapists rarely engage in gratuitous violence, defined as expending energy beyond what is required to subdue or control the victim and inflicting injuries that reduce the victim's chance of surviving to become pregnant or that heighten the risk of eventual injury to the rapist from enraged relatives of the victim (all *ultimate* costs of rape). Thornhill and Thornhill (1992a), having reviewed much of the literature pertaining to men's sexual arousal in response to laboratory depictions of rape, concluded that "not only incarcerated rapists but many other men (the studies collectively implicate young men in general) are sexually aroused to similar degrees by stimuli explicitly portraying consensual sex and rape" (p. 376). This conclusion, however, requires certain qualifications because of recent research. Since the review, a quantitative or meta-analytic review has been published showing that, overall in studies, incarcerated rapists exhibit significantly more sexual arousal in response to depictions of sexual coercion involving physical force than men who have not been convicted of sex offenses (Lalumiére and Quinsey 1994; Hall et al. 1993). But the older literature as well as more recent literature on non-incarcerated men's arousal during exposure to rape depictions indicates that many normal men (college students and community volunteers) are significantly sexually aroused by depictions of coercive sex, including depictions involving physically aggression (Thornhill and Thornhill 1992a,b; Lohr et al. 1997; Proulx et al. 1994). However, men who report a history of sexual coercion show less inhibition in arousal during exposure to forceful depictions than men who do not report such a history (Lohr et al. 1997). Also, as is indicated by some of the laboratory studies reviewed by Thornhill and Thornhill (1992a) and by some more recent research (Malamuth and Linz 1993; Lohr et al. 1997), many men's penile arousal during exposure to depictions of rape is reduced by gratuitous violence toward the victim, by signals of the victim's pain and humiliation, and by the risk to the male of detection of his arousal by coercive stimuli (by requiring men to report their sexual arousal at the same time when arousal is being automatically measured). Thus, there seem to be multiple factors in the degree to which men respond to sexually coercive scenarios, and this implies the possibility of restricting or even eliminating the response. Furthermore, even in anonymous reports only about one-third of men say that they would coerce a woman into sexual acts if they could be assured that they would not suffer any negative consequences (Malamuth 1989; Young and Thiessen 1991). When combined with the laboratory research, this indicates that men's sexual arousal during rape depictions is regulated by factors pertaining to the cost of rape. More research will be needed to determine what specific cues regulate men's sexual arousal to rape and whether or not these cues pertain to rape-specific adaptation. At this time, however, it appears that cues during boys' sexual development and cues that influence adults may affect men's sexual interest in rape. These cues include upbringing without fathers, other correlates of poverty, the nature of relationships in a male's rearing environment, and the presence of an attractive female in a vulnerable circumstance.

Marital rape as a sperm-competition tactic

There also may be rape-specific adaptation related to the rape of one's mate. Only recently, and only in the industrialized West, did men begin to be accused of raping their wives. However, a rape is still a rape, regardless of the victim's relationship to the perpetrator. Rapes and other sexual assaults by husbands, former husbands, boyfriends, and former boyfriends make up about a fourth of all offenses in this category, according to surveys of reported and unreported assaults throughout the United States (Bachman and Saltzman 1995).

Raping an unwilling pair-bonded mate may be a male tactic of sperm competition. A woman's sexual unreceptivity may suggest to her partner that she is having consensual intercourse with another male. Because men associate sexual unwillingness and resistance in their long-term mates with infidelity, sexual unwillingness may lead to sexual jealousy, and sometimes to rape as a sperm-competition tactic. Several types of evidence support this hypothesis. First, the rape of a long-term mate is particularly likely to occur during or after a breakup in which concern about infidelity is directly implicated (Finkelhor and Yllo 1985; Russell 1982). Recent research shows that women separated from their husbands are subject to much higher rates of violent victimization from husbands, former husbands, boyfriends, or former boyfriends than are divorced or married women (Bachman and Saltzman 1995). Second, there is a strong relationship between husbands' wife battering and their sexual jealousy (Daly and Wilson 1988; Jacobson and Gottman 1998). There also appears to be a strong relationship between battering of a wife and rape of a wife. "Sexual matters" were the major source of marital conflict in Finkelhor and Yllo's (1985) study of marital rape.

Russell's (1982) study reveals that sexual matters, including sexual jealousy on the part of the husband, played some role in 53 percent of the beatings reported by victims of wife rape. Of the 137 chronically battered wives studied by Frieze (1980, cited in Thornhill and Thornhill 1992a), 34 percent reported rape as well as battering. Frieze found that the wives of battering husbands who also raped perceived their husbands as more sexually jealous than did wives of husbands who battered but did not rape. This study also reported that the degree and the frequency of violence toward wives by husbands who had both beaten and raped their wives was significantly greater than in cases involving battery without rape. Males in this class also were more likely than others to assault their wives when they were pregnant, and more likely than others to go to extreme lengths to control the behavior of their wives (even to the point of extreme claustration).[16]

A potential cost to a woman of pair bonding for protection is rape by her former protector when the relationship is no longer in the woman's interests and she then leaves or tries to leave. The risk to women of being raped by a current or a past mate may be at least as high as that of being raped by an acquaintance or a stranger (Koss et al. 1987; Bachman and Saltzman 1995).

The Condition Dependence of Rape

Rape by a male of any species, like all traits of an individual organism, has a developmental background, as we explained in chapter 1. Although much remains to be learned about its development, we can be certain that rape is proximately accounted for by gene-environment interactions. This certainty comes from knowledge of how development works. In this sense, rape is condition dependent: It arises from gene-environment interactions that, during development, construct the psychological adaptations that are involved, whether or not these include a rape-specific psychological mechanism. Rape also depends on the interaction of the relevant constructed psychology with environmental stimuli (such as a potential victim who is attractive and vulnerable).

Scorpionfly Rape as a Condition-Dependent Tactic

Another sense in which rape is condition dependent encompasses the sense we have just discussed. This is the sense of a *conditional strategy*.

Rape in scorpionflies is useful for illustrating the evolutionary concept of a conditional strategy (Thornhill 1981, 1984, 1987, 1992a)—i.e., a specialized adaptation with even more specialized tactics as its components. The tactics are adopted by individuals on the basis of particular conditions. Experiments reveal that male scorpionflies prefer to provide mates with nuptial food gifts (dead insects or saliva masses), and that they rape only when they cannot provide such gifts. When some of the males possessing nuptial gifts are experimentally removed, so that their gifts are left unguarded, giftless males (i.e, those unable to obtain dead insects through intense male-male competition for them) quickly shift from attempting rape to guarding an unguarded gift. And males of the species that use both saliva masses and dead insects as nuptial gifts must feed on the latter in order to produce the former. Thus, success or failure in competition for insect carrion is a specific condition that causes a male scorpionfly to adopt a given mating tactic—either rape or resource provisioning. Moreover, a male's phenotypic and genetic quality are conditions that influence his choice of a tactic. The males that employ the rape tactic are smaller or less symmetric than the males that use nuptial gifts. However, experiments reveal that all males will rape in an environment in which nuptial gifts are exceedingly scarce. Studies show that males of high phenotypic quality are more effective rapists than males of low phenotypic quality because they are better at overpowering resisting females.

Male scorpionflies, then, all possess the brain adaptations to employ both resource provisioning and rape as mating tactics and to adopt those tactics adaptively. A male's competitiveness determines his use of these alternative behaviors. Males prefer to provide resources. This is because females prefer males with nuptial gifts and try to avoid rapists. It is also because, both in the evolutionary past and in the present, greater reproductive success is associated with nuptial gift giving. Rape yields much lower male reproductive success than resource provisioning because many rape attempts do not lead to genital union and because only about 50 percent of rape matings result in full insemination. Nuptial feeding results in the male's fully inseminating his mate; indeed, females even compete among themselves to mate with resource-holding males.

Conditional strategies are characterized by marked differences in reproductive success between their tactics. The tactic that yields the least reproductive success involves individuals that are doing the best they can under the limitations they face—making the best of a bad job. Condition-dependent tactical switches are common in organisms because typically a "big winner" alternative will exist that will yield higher reproductive success than other alternative pursuits. The switch from one alternative to another may occur because of conditions encountered during development or in the adult stage, and it may or may not be reversible.

Human Rape as a Conditional Tactic

The hypothesis that human rape reflects rape-specific adaptation views rape as a condition-dependent mating tactic within a single conditional mating strategy of men. Two other behaviors that are tactics of men's mating strategy are honest courtship and deceptive courtship (Shields and Shields 1983). The conditions that affect adoption of rape by men may include limited resource holdings, social disfranchisement, limited sexual access, few rewarding romantic relationships, low phenotypic or genetic quality, and rape opportunities with high benefits and low costs to male reproductive success during human evolutionary history. These circumstances could work as developmental switches that shift males into a lifetime sexual mode involving some or even a great deal of coercion, or as cues that motivate rape in only one sexual encounter. The developmental model of Malamuth (1998) views men as being shifted into sexual coercion by developmental cues of limited resource availability (e.g., father absence) or poor sexual success. The findings of Figueredo et al. (1999)

support a similar developmental model, in that sexual offenders studied were males with reduced psychological and social potency.

The condition dependence of the hypothetical rape adaptation predicts that men may vary in response to rape scenarios in a patterned way. In laboratory experiments such as those suggested above, socially disenfranchised men (i.e., men who fit the model of Figueredo et al.) and men who are more sexually impulsive than others (presumably as a result of limited resources during their rearing—e.g., father absence) are predicted to exhibit greater arousal during exposure to rape depictions. We would add age to the list of conditions. Young men's greater proneness to take risks—itself a product of sexual selection in human evolutionary history—is expected to lead to their being more interested in and more responsive to rape depictions than older men. (Young men have greater libido than older men; however, this could be manifested only in consensual sex, and if it were there would be no age-based difference in response to lab depictions.)

The conditional circumstance of male phenotypic and genetic quality needs further elaboration. As was mentioned in chapter 2, coevolutionary races involving infectious disease organisms and their hosts probably explain why disease resistance and associated health is so highly heritable.[17] This heritability in phenotypic condition is expected to lead to selection favoring individuals that can read their own phenotypic condition and modify their behavior to match it. Psychological adaptation that functions in reading personal condition and adjusting personal tactics accordingly would be the outcome. This is what Figueredo et al. specifically suggest as a model to explain their data on male sex offenders. The offenders are viewed as compensating for genes that interfere with development of psycho-social skills. As a result of a mechanism that assesses the individual's phenotypic condition, the offenders adopt coercive sexuality and criminal behaviors as a tactic for making the best of a bad job. The environmental experiences involved during this kind of development are low performance in social (including sexual) domains of life.

Psychopathy: A genetically distinct rape-related adaptation?

The evolutionary psychologist Linda Mealey (1995) has proposed that men with psychopathy (and thus sociopathy) are a genetically distinct morph, or form, and that normal men don't have the same adaptation.[18]

Psychopathy itself is seen in about 3 percent of men; however, psychopathic tendencies, as measured with a psychometric questionnaire, vary greatly among normal men. This variation may arise from individuals' having received different numbers of the multigenic underpinnings of the psychopathy adaptation at conception. Psychopathy in men is associated with high levels of exploitation of others and with criminal behavior, including rape. The charisma of psychopaths is most effective when people in general are naive—selection favors it only at low frequency. It could be, then, that psychopathic and normal men possess two distinct psychological adaptations with regard to rape—both of which could be condition dependent, in that rape is an output only under certain developmental or ecological circumstances. Additionally, non-psychopathic men would vary in their proneness to rape as a result of the multigenic nature of a psychopathic psychology favorably selected when at low frequency. This is not an argument based on genetic determinism. Receiving genes for psychopathy from parents doesn't inevitably lead to psychopathy. Psychopathy, like every behavioral trait of the individual, is the result of gene-environment interactions during development, and thus it requires external cues for its manifestation (Lykken 1995).

Whether human rape is a result of one or multiple rape adaptations or whether it is all by-product, the condition dependence of rape implies that rape can be reduced—even eliminated—if the conditions affecting it can be fully elucidated.

Does Female Choice Favor Rape in Some Species?

Although there is little if any evidence that rape may be favored by female choice in some species, we must consider the hypothesis in order to clarify certain evolutionary concepts.

Female dung flies (*Sepsis cynipsea*) typically shake and struggle when grasped by a male (Allen and Simmons 1996). The shaking and struggling occur only at this time and in this context. In mating attempts, a male grasps a female with elaborately modified forelegs that clamp the female's wings at their bases and allow the male to hold the struggling female. Struggling females sometimes prevent copulation, as do resisting female scorpionflies and waterstriders. Thus, female struggling acts to select mates that are capable of holding onto them. Allen and Simmons conclude that female struggling, and the resultant rape when their resistance is overcome by certain males, may be a female adaptation that helps females mate with males of superior phenotypic and genetic quality (as demonstrated by their ability to overcome female physical resistance).

It is theoretically possible that female sexual resistance to rapists in some species reflects an evolved mechanism for evaluation of the rapist's "genetic quality" in order to secure genes that will promote the mating success of their sons. Pre-mating resistance on the part of a female may indirectly assess male heritable quality by testing the would-be rapist's strength, endurance, and vigor. If females' resistance results in their mating with males adept at overcoming it, the sons of rapists will be similarly adept, having inherited the genes of their fathers. These ideas, first put forth by the biologists Kathleen Cox and Burney Le Boeuf (1977) as a hypothesis to explain female resistance to mating in certain seal species, were subsequently discussed by Thornhill (1980), Thornhill and Alcock (1983), Arnqvist (1992), Eberhard (1996), and Allen and Simmons (1996) in relation to female sexual resistance in certain non-human species. The biologists Robin Baker and Mark Bellis (1995) extended the hypothesis to female physical resistance during rape in humans, which, they suggest might act as selection on would-be rapists, with the consequence that females will be inseminated only by males able to overcome the victim's resistance. As in the case of the dung fly, it is conceivable that in the past women who filtered potential rapists by resisting them bore sons who turned out to be adept at raping and thus may have had more grandchildren than passive females.

To determine whether resistance is functionally designed as female choice for adept rapists, whether it is an adaptation to avoid rapists altogether, or whether it is an adaptation to minimize the costs of high mate number, more research is needed on species in which females show physical resistance to mating. However, female physical resistance for the sake of mating with a male of superior phenotypic and genetic quality appears unlikely to evolve commonly, and there is no evidence of such an adaptation in human females. There are many easier and less costly ways for females to gain phenotypic and genetic information about males from males' non-coercive signals and from the outcomes of male-male agonistic interactions. Also, costs to the female in the context of physical resistance would include mating errors resulting from many females' incapability of resisting unwanted mates. For these reasons, we interpret human rape as a circumvention of female choice, and we interpret the act of rape as having consistently reduced female reproductive success during human evolutionary history.

But what if there was evidence that human rape was an adaptation that had been selected because it increased the reproductive success of females as a result of the high mating success of their rapist sons? Would that imply that rape was "natural" and therefore good? Would it imply that rape was something females ought to enjoy and encourage because it had increased the reproductive success of females in ancestral populations? Would it imply that feminists should celebrate rape as a form of female power? We think not. To think otherwise is to fall prey to the naturalistic fallacy.

Summary

Although the question whether rape is an adaptation or a by-product cannot yet be definitively answered, the evolutionary approach illuminates many aspects of why men rape. The ultimate causes of human rape are clearly to be found in the distinctive evolution of male and female sexuality. The evidence demonstrates that rape has evolved as a response to the evolved psychological mechanisms regulating female sexuality, which enabled women to discriminate among potential sex partners. If human females had been selected to be willing to mate with any male under any circumstances, rape would not occur. On the other hand, if human males had been selected to be sexually attracted to only certain females under certain limited circumstances, rape would be far less frequent. Indeed, if human males had been selected to desire sexual intercourse only with females who showed unmistakable willingness to copulate with them, rape would be an impossibility. Human rape exists because selection did not favor these types of adaptations, and the proximate causes of human rape lie in the different adaptations of male and female sexuality that were formed by selection in human evolutionary history. Hence, the sexual adaptations that exist in men and women, described in chapter 2, provide the best guide to creating environments that will decrease the frequency of rape.

4

The Pain and Anguish of Rape

Evolutionary theory can help us understand the ultimate reasons why rape is as distressing as it is[1] by focusing on how it interfered with reproductive success in ancestral human populations. This interference took three main forms: it reduced the victim's fitness by circumventing mate choice, it reduced the fitness of her mate by lowering his paternity certainty, and it reduced the fitness of the relatives of the victim and her mate as a result of the preceding two factors.

Psychological Pain

By *psychological pain* we mean the mental state of feeling distraught. (All pain is psychological in the sense that it exists in the brain.) Thornhill and Thornhill (1989) hypothesized that the human capacity to experience psychological anguish is an adaptation that functions to guide cognition, feelings, and behavior toward solutions to personal problems that reduced individual reproductive success in human evolutionary history. In essence, they hypothesized that psychological pain is an adaptation that functions against such losses by focusing attention on the causes of the losses. The result is that attention is directed toward ways of dealing with current circumstances, given the loss, and of avoiding a repetition of the events that caused the loss. Psychological pain, like any feeling, is generated by psychological adaptation.

This evolutionary hypothesis about psychological pain makes two general predictions about the kinds of environmental information that will cue or activate psychological pain:

- These cues will be events that lowered reproductive success in human evolutionary history.
- The greater the negative effect of an event, the greater the psychological pain experienced.

Loss of social status, death of a relative, desertion by a mate, and being raped are examples of events predicted to generate great psychological pain. Variation in degree of psychological pain can be predicted in each of these categories of loss. For example, death of a high-reproductive-potential relative is expected to result in more pain than death of a low-reproductive-potential relative, and in the event of rape more psychological pain is expected in a young and fertile woman than in a female of pre- or post-reproductive age. A graded expression of psychological pain is expected because psychological pain has reproductive costs as well as potential benefits. For example, psychological pain can distract a person from many important matters of everyday life. Hence, the psychological mechanisms governing psychological pain should calibrate the degree of pain to the benefits of solving the problem and avoiding repetition of the painful experience measured against such costs.

Rape Victimization in Evolutionary Context

In human evolutionary history, rape could have resulted in a reduction in female reproductive success in the following four ways:

- The victim may have been physically injured.
- Rape reduces a woman's ability to choose the timing and circumstances for reproduction, as well as her ability to choose the man who fathers her offspring. Thus, when rape led to conception and to production of the rapist's child, the woman may have expended her limited parental effort on an offspring of lower genetic quality than one she could have produced with a partner of her choosing.
- Rape circumvented a woman's ability to use copulation as a means of securing material benefits from men for herself and/or for her offspring.
- Rape of a pair-bonded woman may have adversely influenced her pair-bond mate's protection of her or the quantity and quality of parental care her offspring received from him.

Human males are unusual among mammals in their potential for parental investment. Parental care from both sexes has been crucial to human reproductive success (Alexander and Noonan 1979; Benshoof and Thornhill 1979; Lancaster 1997). But men care more for their genetic offspring than for the offspring of others (Daly and Wilson 1988; Wilson and Daly 1992). Actual or suspected rape of a man's mate sometimes reduces his confidence that he sired the mate's previous offspring and his confidence that he will be the sire of the next offspring if his mate becomes pregnant around the time of the rape. As was discussed in chapter 2, in human evolutionary history this perception could have negatively influenced a male's behavior toward a female and toward her offspring. If so, being raped decreased the victim's reproductive success. Even attempted rape may have been of great concern to our male ancestors from the standpoint of paternity reliability, since in the paternity-focused, adaptively paranoid male mind a woman who placed herself in a situation conducive to a rape attempt may fail to avoid similar situations in the future.

On evolutionary theoretical grounds, men's concern about the rape of their mates is expected to be specific to rape by other humans. That men's concern about rape's lowering their confidence of paternity has this specificity is suggested, albeit anecdotally, by an instance in which a woman was raped by a male orangutan. Male orangutans often rape female orangutans in the wild (Wrangham and Peterson 1996). The orangutan involved in this particular rape had been born in the wild and captured for research purposes. He was relatively tame around the humans at the jungle camp of the research group. However, one day he attacked and raped a cook at the camp despite attempts by the veteran orangutan researcher Biruté Galdikas to stop him. Wrangham and Peterson (ibid., pp. 137–138) summarize Galdikas's comments on what transpired after the rape: "Fortunately, the victim was neither seriously injured nor stigmatized. Her friends remained tolerant and supportive. Her husband reasoned that since the rapist was not human, the rape should not provoke shame or rage." Galdikas (1995, p. 294) recalls the husband saying: "Why should my wife or I be concerned? It was not a man." Neither the husband nor the victim seemed to suffer greatly.

We feel that the woman's perspective on rape can be best understood by considering the negative influences of rape on female reproductive success—especially the last three of the four factors listed above, which would have reduced a woman's options for maximal reproductive success in the evolutionary environment of human history. If rape was an event that often negatively affected female reproductive success in human evolutionary history, rape victims are expected to experience psychological pain because, in the past, females with similar psychological mechanisms were motivated to focus attention on the circumstances that resulted in the pain and to avoid them in the future. It is also highly possible that selection favored the outward manifestations of psychological pain because it communicated the female's strong negative attitude about the rapist to her husband and/or her relatives.

The evolutionary perspective on psychological pain predicts that, in general, such pain will be manifested in women who are victims of rape, and indeed this is quite obviously true.[2] The evolutionary perspective also makes specific predictions about the characteristics of rape victims that will influence the degree of their psychological pain. Before outlining these specific predictions, we will briefly discuss the data set.

The data, obtained from the Joseph Peters Institute in Philadelphia, involve 265 variables and 790 rape victims. The victims were females of all ages who reported to authorities a sexual assault (primarily rape) and who were examined at the Philadelphia General Hospital between April 1, 1973 and June 30, 1974. (Victims of age 12 and younger were included through June 30, 1975.) The victims were interviewed by social workers within five days after the rape. When the victim was a child, a caretaker sometimes helped the child respond to interview questions; when the victim was very young, the caretaker responded to the questions on the basis of his or her perception of the effect of the sexual assault on the child.

The results discussed here pertain to twelve variables associated with psychological pain. These variables were used by the researchers at the Joseph Peters Institute to measure the psychological adjustment of the victims on the basis of a recording of each victim's verbal report. Thus, these variables measure the magnitude of the psychological pain experienced by each victim. The raped females were asked to evaluate fear of being on the street alone, fear of being home alone, change in social activities, change in eating habits, change in sleeping habits, frequency of nightmares, change in non-sexual relationships with men, change in feelings toward known men, change in feelings toward unknown men, change in relations with husband or boyfriend, change in sexual relations with partner, and insecurities concerning sexual attractiveness. The responses indicate the impact of rape on each victim's ability to cope psychologically with each of the twelve circumstances. Pre-reproductive-age females were not asked about sexual matters or mateships.

The 790 victims in the sample ranged in age from 2 months to 88 years. The mean age was 19.6 years, the mode was 16 years, and 19 percent were children (0–11 years). Victims of age 12 and younger were included for an additional year after adults were no longer included because the researchers were trying to boost the sample of child victims (McCahill et al. 1979). As we have discussed, rape victims are usually young women of high fertility, with an average age in the early twenties.

At the time of the incident, 81 percent of the victims were widowed or divorced or had never married, approximately half were receiving some form of public financial assistance, and almost all (725 of the 790) had an annual income of $12,000 or less. For 80 percent of the victims, this rape was the first sexual assault ever experienced; 13 percent had experienced one previous sexual assault, and the remainder had experienced two or more.

Although it is a large sample of rape victims, the sample is not representative of females in the United States. It contains a disproportionately large percentage of very young and unmarried females of low socioeconomic standing. It also includes only females who reported the assault to the authorities. According to recent estimates, only about 16–33 percent of rapes are reported to police (Kilpatrick et al. 1992; Greenfield 1997). However, there is no reason to believe that the sample is atypical with respect to the psychology of psychological pain being analyzed here.[3]

Predictions and Findings

Age of Victim

Young women (i.e., females of reproductive age) are predicted to suffer greater psychological distress from rape than girls (non-reproductive-age females) or older women (of non-reproductive age) (Thornhill and

Thornhill 1983). Again, this prediction derives from the correlation between age and conception risk and thus from rape's potential to circumvent female choice. Only reproductive-age females can become pregnant in the event of rape. The psychological-pain hypothesis predicts that psychological pain will be experienced in direct relation to the effects that the pain-causing incident would have had on individual reproductive success in the human evolutionary historical environment.

In the study that examined this prediction empirically (Thornhill and Thornhill 1990a), victims' ages were divided into two categories: "non-reproductive" (1–11 and 45–88 years) and "reproductive" (12–44 years old).[4] The study showed that reproductive-age rape victims suffered significantly more psychological trauma than non-reproductive-age rape victims.

It was important to examine whether the significant difference between reproductive-age women and non-reproductive-age females in psychological pain was a result of either pre-reproductive-age girls' or post-reproductive-age women's having little psychological trauma relative to reproductive-age women. For this purpose, the sample was recategorized by age into three categories: pre-reproductive (0–11 years), reproductive (12–44), and post-reproductive (45+), and the analysis was then rerun. This analysis showed that reproductive-age rape victims experienced the greatest psychological pain and pre-reproductive rape victims the least. That the psychological trauma of pre-reproductive rape victims is relatively low has also been suggested by findings from other studies, discussed in Thornhill and Thornhill 1990a.

Mateship Status

Here the prediction was that rape would adversely influence the victims' relations with their husbands, thereby leading to psychological pain in victims (Thornhill and Thornhill 1983). This prediction stems from the negative effect of rape on paternity reliability and the consequent lowered potential for successful reproduction by a pair-bonded man via his mate. According to the prediction, the victims' pain stems from their mates' reduction or complete withdrawal of material support from them and their children. And, indeed, married women were found to be more traumatized by rape than unmarried women (Thornhill and Thornhill 1990a).

Married women are more likely to be of reproductive age, and perhaps that is why they show greater psychological pain after rape. However, when the effect of marriage on post-rape psychological pain was controlled for, reproductive-age females were still found to exhibit significantly more pain (ibid.). This suggests that age itself is a proximate cause of psychological distress after rape.

When asked whether they felt that the rape would affect their future, both reproductive-age victims and married victims were more likely to feel that their future was harmed by the rape than non-reproductive-age victims (of either category) and unmarried victims, respectively (ibid.). This pattern is as predicted: in evolutionary terms, rape would have been more costly to women of reproductive age, married or not.

Overall, then, this analysis indicates that age and marital status are proximate causes of the extent of psychological pain after rape. That is, the results indicate that the female psychology that regulates this part of psychological pain is affected by the age and the mateship status of the rape victim (Thornhill and Thornhill 1990a).

Additional analysis of the data (Thornhill and Thornhill 1990b) showed that strangers, rather than friends and family members, were the most frequent perpetrators in the sample, and this was more the case among the adult women than among the girls. This analysis also revealed that, of rape by friends, by family members, and by strangers, the last was associated with the greatest psychological pain.[5] However, reproductive-age victims and married victims experienced more psychological pain than other categories of victims, regardless of whether the rapist was a stranger, a friend, or a family member. These victims did not exhibit more psychological pain simply because they tended to be raped primarily by strangers.

Force and Violence

Another potentially confounding variable in the patterns of age and mateship discussed above is the amount of force and violence used by the perpetrator. Reproductive-age rape victims were more often subjected to violent attacks than victims in the other two categories (Thornhill and Thornhill 1990c). This result was predicted on the basis of the combination of two factors: that reproductive-age females should be most likely to fight back and even to escalate their resistance because of the greater evolutionary historical cost to their reproductive success of being raped, and that rapists would be more highly sexually motivated to rape reproductive-age females and would be more willing to incur a higher cost (possibility of injury) in these rapes because of the victim's greater sexual attractiveness to men relative to females in the other two categories.

That greater force and violence were used in raping reproductive-age females does not confound the correlation between victim's age and psychological pain. Even in the absence of force or violence, reproductive-age victims were still significantly more psychologically traumatized than either pre- or post-reproductive-age victims. There was no significant difference in the amount of force and violence endured by married versus unmarried victims. Thus, a difference in amount of force and violence between rapes of women in the two marital categories does not confound the pattern of greater psychological trauma of married rape victims (Thornhill and Thornhill 1990c).

McCahill et al. (1979) performed an initial analysis on this same data set and reported that violence accompanying a rape showed an overall negative relationship with psychological distress after the rape. That is, as the violence increased, the psychological pain of the victim declined. This pattern surprised McCahill's research team; they expected more violence to lead to greater psychological trauma in victims. Of course, their pattern may have been confounded by the various variables we have discussed. It was predicted, however, that the occurrence of violence during rape would have a moderating effect on the psychological trauma of reproductive-age victims, particularly those in significant mateships (Thornhill and Thornhill 1983). This was predicted because a victim may have less difficulty convincing her mate that rape rather than consensual sex has occurred if she exhibits physical evidence that sexual access was forced. A single rape is less threatening to paternity than frequent consensual sex. If blame and incredulity on the part of the victim's mate are proximate causes of psychological trauma after rape, then victims are expected to experience less psychological pain when their mates are less blaming and less incredulous. This is also consistent with the view that outward manifestations of psychological pain communicate to the victim's mate that she really was raped.

Data on victims' self-reports of force and violence and independent evidence of violence taken by medical examiners make it possible to test whether reproductive-age victims who had not experienced violence exhibited more psychological pain than victims of the same age who had been attacked violently. In fact, this has been shown (Thornhill and Thornhill 1990c). As also predicted, Thornhill and Thornhill found that the married women whose rapes had been marked by violence exhibited less psychological pain. Thus, reproductive-age married women appear to be less psychologically traumatized when the rape includes violence, thus providing clear evidence to their husbands that copulation was not consensual. This result supports the earlier findings that mateship status is a proximate cause of the psychological pain experienced by rape victims. Also, it indicates that, in addition to the victim's age and mateship status, a third proximate cause—the credibility of the rape report to the pair-bond mate—is involved in women's psychological pain after rape.

Sexual Behaviors

In a final analysis of this data set (Thornhill and Thornhill 1991), the influence of the nature of the sexual behavior on the victim's psychological pain was examined. The type of sexual behavior involved in sexual assaults should be related to the man's sexual motivation. If men have evolved sexual preferences for fertile women, then women of reproductive age—relative to pre- or post-reproductive-age females—should more often be victims of sexual assaults that include penile-vaginal intercourse, ejaculation in the victim's reproductive tract, and repeated intercourse. In general, all three of these predictions are supported by the data (Thornhill and Thornhill 1991). These three predictions complement the prediction that reproductive-age women should be highly overrepresented among female rape victims—something that now has been shown repeatedly in a variety of wartime and peacetime settings. It appears, then, that men prefer to rape young women, and that when they rape such women they are more strongly sexually motivated than when they rape pre- or post-reproductive-age females.

The view that men who rape are sexually motivated is also supported by the evidence that men are more likely to rape reproductive-age females even though they may resist more.

The finding that sexually assaulted non-reproductive-age females are less likely than reproductive-age females to experience penile-vaginal intercourse or multiple copulations or to receive the rapist's ejaculate vaginally does not imply that rapists of non-reproductive-age females lack sexual motivation entirely. Single penile-vaginal copulations and assaults involving anal intercourse, fellatio or cunnilingus, and other forced touching of the female's genitals depend proximately on a perpetrator's sexual interest.

The difference in the sexual behaviors experienced by reproductive-age victims during sexual assaults is not the reason that such victims have greater psychological pain (Thornhill and Thornhill 1991). Pre-reproductive-age girls who were victims of penile-vaginal intercourse were no more psychologically traumatized than pre-reproductive-age girls who were not. The same appeared to be true of post-reproductive-age women. Yet reproductive-age women who were victims of penile-vaginal copulation reported more psychological pain than those who were victims of other sexual behaviors (e.g., fellatio, cunnilingus, anal intercourse) without penile-vaginal intercourse. It is only in reproductive-age females that penile-vaginal intercourse is associated with an increase in psychological pain after a sexual attack.

There was also evidence that reproductive-age females, but not infertile victims, suffer more psychological pain when sperm is ejaculated inside them during rape than when it is not. This effect of ejaculation was smaller than that of penile-vaginal intercourse, and there was no effect of repeated episodes of copulation on the psychological pain of reproductive-age victims (Thornhill and Thornhill 1991).

Thus, as predicted, degree of psychological pain closely follows the likelihood that rape circumvents female mate choice. Since copulation is highly correlated with ejaculation in men (Symons 1979), victims' awareness of sperm receipt and multiple copulation may be less reliable indicators of risk of fertilization than their cognizance of copulation.

Proximate Causes of Psychological Pain after Rape: An Overview

The results presented in a series of articles by Thornhill and Thornhill (1990a–c, 1991) indicate that the psychological mechanisms that regulate psychological pain in response to rape are affected by a woman's age (and thus her fertility), by her marital status, by her treatment by the rapist, and by whether there was penile penetration of the vagina. Over human evolutionary history, each of these factors would have affected the likelihood of unwanted pregnancy and potential loss of a mate's protection, resources, and paternal care. These findings are consistent with the hypothesis that human rape has been a sufficient obstacle to the reproductive success of adult females to have led to psychological mechanisms designed by natural selection to increase the ability of women to avoid future rapes.

In recent years, evolutionarily informed research on human unhappiness has increased considerably, and there is growing evidence that psychological pain is an adaptation that defends against the circumstances that reduced the reproductive success of individuals in human evolutionary history.[6] The biologists Paul Watson and Paul Andrews believe that depression is functionally designed for assessment of the cues relevant to the evolutionary psychological pain hypothesis (unpublished manuscript). Their review reveals that during psychological pain individuals are almost entirely focused on the problems causing the pain, that they have more objectivity about themselves and their status in their social network, and that they possess enhanced problem-solving skills.

Future Research on Psychological Pain of Rape Victims

To our knowledge, no research reported after 1991 has examined the psychological pain of rape victims from a Darwinian perspective.[7]

Future research on psychological pain associated with rape would be facilitated if it were possible to distinguish between unmarried rape victims with boyfriends and those without boyfriends. Unmarried victims with investing boyfriends may respond more like married victims, since in both cases cuckoldry is possible in the event of rape.

We realize that psychological pain is a complex and multifaceted mental state that may include a wide diversity of negative feelings—even so-called somatic (bodily) pain. It may be possible to derive predictions from the psychological-pain hypothesis about the mix of the various negative emotions (anxiety, fear, sadness, anger, guilt, shame) expressed by reproductive-age women versus other age categories of victims as time elapses after the rape and in different settings (e.g., pair-bonded or not).

Future research might also examine behaviors of rape victims that are expected under the psychological-pain hypothesis. For example, victims paired to investing males are predicted to emphasize their resistance when recounting rapes to their partners, especially in the absence of physical evidence of force. In addition, the nature of changes in victims' social activities after rape should be predictably based on age and mateship status. For example, reproductive-age victims may show more fear of strange men and unfamiliar social settings after rape than victims in other age categories, and the fear may be specific to circumstances related to the potential of the occurrence of rape. Moreover, victims taking psychotropic drugs to alleviate psychological pain after being raped might be compared with victims not taking such drugs in order to determine if the drug users experience disadvantages in coping psychologically with rape-related problems. If psychological pain is an adaptation that aids in solving problems, eliminating the psychological pain by medication is predicted to lead to a longer period of pain and perhaps even to an inability to resolve the problem and/or avoid its recurrence.

Finally, we predict a significant sex difference in factors related to psychological pain after rape. Male rape victims also report psychological pain (Rogers 1995). Little is known about rape-related psychological pain in men because only recently has it received any attention. We expect that studies will find that male victims' pain is related primarily to perception of loss of status. A raped male is socially impotent when it comes to influencing the behavior of the perpetrator(s) and may be viewed by others as socially impotent in general. Because of the strong, positive relationship between status and reproductive success of males in human evolutionary history, loss of status is anticipated to stimulate psychological pain in male humans (Thornhill and Thornhill 1989). We expect, also, that age of victim will not predict psychological pain after rape in men.

The findings on the pain of rape victims indicate that women have a special-purpose psychological adaptation that processes information about events that, over human evolutionary history, would have resulted in reduced reproductive success. However, more research is needed to demonstrate the specificity of such information processing in cases of rape relative to other crimes against women. For example, if young women are more psychologically traumatized than older women after theft of property without physical contact or threat of it, the greater psychological trauma of young women rape victims would not be specific to rape. The evolutionary psychological-pain hypothesis predicts that women's psychological pain in the event of such theft will correlate positively with the value of the property and not with the victim's age.

The evolutionary approach to the study of rape victims' psychological pain holds great promise of discovering the detailed nature of the cues that activate the pain and, therefore, of alleviating their suffering.

Female Adaptation against Rape across Species

Research on rape victims' psychological pain is grounded in the aspect of evolutionary theory that deals with coevolutionary contests between males and females when their reproductive interests are not the same. The theory of intersex conflicts over when or whether mating will occur predicts the evolution of female counter-adaptations to male traits for coercive sex that increase the reproductive success of males but reduce that of females (Parker 1979; Eberhard 1985; Clutton-Brock and Parker 1995). Sexual coercion is costly to females in many animal species (Mesnick 1997). Coercive males may force females to spend time and energy avoiding them, circumvent females' mate choice, thwart females' freedom of movement, disrupt females' feeding or maternal care, cause females' partners to abandon them, disrupt females' reproductive cycles, and/or cause abortions by rape.[8] Indeed, as happens among humans, in some bird and non-human mammalian species females sometimes die or are seriously injured as a direct result of rape.

Across species, females' counter-adaptations to rape are diverse. They include forming mating alliances with males who guard against sexual coercion by other males, mating with males for convenience (to avoid harassment, injury, or loss of time), forming alliances with non-mate male "friends," forming female-female coalitions, avoidance, resistance, physiological counters (including anti-insemination and abortion mechanisms), and body projections that reduce rapists' sexual access (Mesnick 1997; Smuts and Smuts 1993; Clutton-Brock and Parker 1995; Gowaty and Buschhaus 1998).

The biologists Göran Arnqvist and Locke Rowe's (1995) studies of waterstriders may be the most detailed research that has been done on a female adaptation against rape in a non-human species. As has already been mentioned, male waterstriders have a special-purpose rape adaptation: a pair of ventral abdominal projections with which they can hold onto females that resist copulation. Copulation is costly to female waterstriders because both the male and the female are more vulnerable to fish predators while it is in progress and because its restriction of a copulating female's movements limits her ability to escape predators and to feed (Rowe et al. 1994). Female waterstriders also have a pair of abdominal spines—dorsal rather than ventral—that are an anti-rape adaptation. Arnqvist and Rowe (1995) observed that female waterstriders that had longer dorsal spines mated less frequently than those with shorter spines. They got the same result by experimentally altering females' spine lengths. Female waterstriders with longer spines could more effectively thwart mating attempts of harassing males. In addition to their dorsal spines, the female waterstrider appears to have a special-purpose rape-avoidance behavior: She can perform spectacular somersaults when an unwanted male grasps her, and these maneuvers often dislodge the male. The fact that this avoidance behavior is energetically very expensive emphasizes the importance of mate choice to female waterstriders (Watson et al. 1998).

Female scorpionflies have a suite of behaviors that appear to function in resisting forced copulation. As was noted in chapter 3, female scorpionflies prefer males that offer them nuptial food gifts in exchange for mating. Males experimentally given food that they are able to use for this purpose were selected as mates. When the same males were prevented from offering food, they attempted (sometimes successfully) to force copulation. But during rape attempts, females made strong and often successful efforts to escape (Thornhill 1980, 1984; Thornhill and Sauer 1991). Even when grasped in genital contact, female scorpionflies can sometimes prevent insemination by rapists. In one experiment (Thornhill 1984), only half of rape copulations by male scorpionflies led to sperm transfer, whereas males bearing nuptial gifts inseminated the now-willing females 100 percent of the time.

Another rape counter-strategy used by female scorpionflies is a quick return to sexual receptivity after rape with insemination. Apparently the receptivity of females is controlled by chemicals in the male's ejaculate, as is known to be the case in some other insects. The chemical is an adaptation against sperm competition. When a female scorpionfly is raped and inseminated, she returns to sexual receptivity twice as fast as after consensual copulation. This allows a female to choose a mate bearing a nuptial gift soon after a rape (Thornhill 1980, 1984).

Ejaculate dumping—which occurs in a variety of animal species, including insects, birds, and mammals (Eberhard 1996)—is another female tactic that may reduce the chances of being impregnated by a coercive male. For example, in the red jungle fowl *Gallus gallus* (the wild ancestor of the domestic chicken), females prefer socially dominant males as mates. Subordinate males rape. A rape attempt sometimes involves a long chase, during which the female tries to get a dominant male to disrupt it. But should a subordinate male succeed in inseminating an unwilling female, the female often immediately expels a portion of the rapist's ejaculate.[9] Ejaculate dumping promotes a kind of female choice that is known as *cryptic* (Thornhill 1983; Eberhard 1996) because it occurs during or after mating rather than before mating.

A woman may reject much of a rapist's sperm via the copious "flowback" typically associated with the absence of orgasm during mating (Baker and Bellis 1995). Human rape victims rarely show much sexual arousal and almost never achieve orgasm. It is conceivable that some aspects of women's capacity for orgasm evolved in the context of reducing the fertilizing capacity of rapists' ejaculates. That is, the absence of orgasm during rape may be an evolved response to rape.

Men's Evolutionary Counter-Strategies

That some rapes result in pregnancies suggests that men may have evolved counter-strategies to the rape-fertilization-control defenses women may possess. The clearest evidence for this is the relatively high rate of fertilization through rape during warfare. During Rwanda's recent civil war, as many as 35 percent of 304 rape victims surveyed may have became pregnant (McKinley 1996), and a high percentage of the rape conceptions resulted in offspring despite the fact that most of the women claimed not to want the pregnancies.

Estimates of the rates of pregnancy resulting from rape in peacetime settings vary from 1 percent to 33 percent. The highest estimate is from a study of pregnant teenagers, 33 percent of whom reported experiencing forced or unwanted sexual intercourse (Moore 1996). The most convincing study of pregnancy and rape in peacetime settings (Holmes et al. 1996) involved a three-year longitudinal study of a representative sample of several thousand American women. Among victims of reproductive age (12–45), the rape-related pregnancy rate was 5 percent per rape, or 6 percent per victim. Of these pregnancies, 38 percent led to birth (the child being either kept by mother or put up for adoption); 12 percent resulted in spontaneous abortion; 50 percent were terminated through clinical abortion. However, 5 percent is probably an overestimate of the rate of pregnancy from rape. Hammond et al. (1995), whose work was based on *in utero* DNA paternity tests, report that 60 percent of women who become pregnant after rape were impregnated by a consensual mate. Thus, the figure reported by Holmes et al. probably should be corrected to about 2 percent. At this time it is not known whether false rape allegations influence this percentage.

The paternity study by Hammond et al. indicates that consensual mates' ejaculates are competitive with ejaculates of rapists in fertilizing eggs. Among certain ducks and among red jungle fowl, a male typically copulates with his partner immediately after discovering that she has been raped by another male. To our knowledge, it has not been determined whether men's ejaculate size or their sexual interest in a mate increases immediately after discovery that the mate has been raped.

Women's Rape Avoidance

As we have noted, women appear to have a psychological adaptation involving psychological pain that comes into play after rape. But evolutionary theory would also predict that women possess psychological adaptations designed to prevent rape from ever occurring, perhaps by motivating avoidance of significantly risky situations. This hypothetical adaptation would adjust a woman's anxiety and fear in accordance to her vulnerability to rape (Thornhill 1997b). Such a prediction is supported by the results of a study conducted in Christchurch, New Zealand (Pawson and Banks 1993).[10] Young (more fertile) women were found to be more fearful of assault while in or outside their homes than older women, and young women's fear was found to be more focused on sexual assault whereas that of older women was more focused on burglary. Also, the amount of a young woman's fear corresponded to the likelihood of actual police-recorded rapes across different sections of the city. (That women in general fear rape and that this fear influences and limits their behavior [e.g., their nocturnal activity patterns] is well known; see, e.g., Riger and Gordon 1981.)

One could study this further by exposing women to slides or narratives depicting environments in which rape is expected to be more or less likely and asking for their emotional responses to these settings. We predict that young women, especially those in the follicular (fertile) aspect of the menstrual cycle, would be most astute at evaluating the dangers of various environments. These individuals should be especially good at distinguishing differences related to the probability of rape, such as how socially isolated they are and whether young or socially disenfranchised men are present.

Chavanne and Gallup (1998) found that young women exhibited a considerable decrease in behaviors associated with risk of rape (e.g., walking in a dimly lit area) when they were not "on the pill" (and thus had ovulatory cycles) and when they were in the fertile phase of the menstrual cycle rather than in the infertile phases. That study involved a large sample of undergraduate women with a mean age of 22. The women completed an anonymous questionnaire that included a range of 18 activities that they could have been involved in during the past day. The questionnaire also requested information about where they were in their menstrual cycle and what method of contraception they were using. The 18 activities were separately rated by a different group of women on the basis of the extent to which each activity would make someone vulnerable to sexual assault. The 40 raters showed similar ratings of the 18 activities in terms of risk of sexual assault. Among the women taking birth-control pills, there was little variation in risk taking in relation to menstrual-cycle phase. (The variation was not even close to statistical significance.) Women not "on the pill," however, show a significant decrease in risk taking during the ovulatory phase of the cycle (days 13–17), and only during that phase. Rogel (1976) found that, among about 800 victims of sexual assault, proportionately fewer women were raped during the middle portion of the cycle, and this was especially the case for women who were in their late teens and early twenties. Morgan (1981) got a similar result using a smaller but still substantial sample (123) of female sexual-assault victims.

The reduction in risk taking by women of peak fertility (ovulatory, mid-cycle) may be specific to rape. There is growing evidence that women move around more and make greater efforts to mate during the ovulatory phase of the cycle than during other phases. In a study using pedometers, Morris and Udry (1970) found that women walked more at mid-cycle than during other phases.[11] Sexual activity also appears to increase at mid-cycle (Chavanne and Gallup 1998; Gangestad and Thornhill 1998; Thornhill and Gangestad 1999). Women exhibit changes in sexual preferences across the cycle. The findings that they prefer masculine facial features and the scent of symmetric men at mid-cycle imply that they pursue male genetic quality during the time of highest probability of conception. Also, women's increased mating effort at mid-cycle results in more extra-pair copulation in this phase of the cycle than in others (Baker and Bellis 1995). Thus, women are most active and most sexual at mid-cycle; however, in that phase they appear to specifically avoid activities that would put them at risk of rape.

Perhaps women exhibit less activity related to risk of rape at mid-cycle owing to their compliance with their mates' paternity-related increase in control of their behavior at that phase. There is no strong evidence, however, that men can detect that their mates are in the ovulatory phase (Benshoof and Thornhill 1979; Burley 1979; Baker and Bellis 1995). Recent studies indicate that the olfactory attractiveness of women to men is unrelated to cycle phase or to use of oral contraceptives (Thornhill and Gangestad 1999). Thus, the behavior that Chavanne and Gallup recorded appears to be attributable to the female alone, without her mate's influence.

As the above discussion illustrates, numerous predictions about women's rape-avoidance behavior can be drawn from evolutionary theory. Some are general, such as that young women will be more adept than post-reproductive-age women at detecting rape-relevant cues. Some are more specific, such as the influence of menstrual-cycle phase on assessment of rape risks. And we can derive others that are highly specific—for example, that among women of the same age peer-rated physical attractiveness will correlate positively with ability to detect rape risks, and even that a woman's waist size will be related to her detection of rape risks.

The idea that women have evolved to avoid rape also may help explain certain aspects of the feminist movement, since opposition to sexual coercion of all forms—but especially rape—is a major concern of that movement. As women achieve more social independence and greater economic opportunity, they are likely to come into more contact with men who are little known or entirely unknown to them. Living away from her natal family, as more women now do, reduces a woman's protection by her male kin (Smuts 1992; Geary 1998). We suggest that the combination of greater mobility and less protection by mates and male kin results in women perceiving an enhanced risk of sexual coercion. This perception (probably accurate) may have fueled the feminist movement's promotion of the kind of female-female alliances against male coercion that are seen in many other mammalian species.[12]

Summary

Human females and females of many other species show evidence of adaptation against rape. Women's psychological pain after rape—a fairly well-studied trait—appears to be an adaptation that defends against events that, during human evolutionary history, resulted in reduced reproductive success of rape victims. Evidence for this derives from the fact that the hypothesis that rape victims' psychological pain is a function of reduced reproductive success caused by a rape predicts the correlates of psychological pain of sexual-assault victims. A rape victim's degree of psychological pain depends on her age (more psychological pain if she is of reproductive age), on her mateship status (more psychological pain if she is married), on the nature of the sex act (more psychological pain if penile-vaginal intercourse occurred), and on whether or not there is evidence that copulation took place without her consent (more psychological pain if there is no physical evidence of resistance). Rape reduces mate choice to the greatest extent in fertile adult females because only they can get pregnant from rape. A rape victim may lose her mate's support, since men value paternity reliability in relationships in which they invest parental effort.

There is also some evidence that women may have psychological mechanisms designed to reduce the occurrence of rape. Young women are more fearful of rape, and young women in the most fertile phase of the menstrual cycle seem to avoid environments with high rape risk.

The ability of women to reject the ejaculates of rapists, if it exists, is clearly incomplete, since some rapes do result in pregnancy.

5

Why Have Social Scientists Failed to Darwinize?

We are proud to present this monograph . . . exposing the application of genetic determinism to justify racist and sexist theories and activities.
—Tobach and Rosoff 1985, p. v

Why have attempts to explain rape by means of the only scientific ultimate explanation of living things been greeted with charges of racism and sexism? Why, at universities, have women's study groups tried to prevent public lectures on the evolutionary basis of rape, and why have picketing and audience protests caused such lectures to be cancelled or terminated? Why have researchers attempting to discover the evolutionary causes of rape been denied positions at universities? Why have organizers of scholarly conferences attempted to keep papers on evolutionary analysis of rape from being presented? Why have editors of scholarly journals refused to publish papers treating rape in a Darwinian perspective? Why have administrators of rape crisis centers refused to assist in research on rape victims' psychological pain upon discovery that the research was based in biological theory?

The answer to these questions can be found in the answer to the more general question of why so much misunderstanding of evolutionary theory continues to exist within the social sciences.

The Historical Neglect of Adaptation

For various reasons, evolutionary studies in biology after Darwin and until the 1960s focused primarily on speciation and on microevolution. 'Speciation' refers to the processes—including selection—that are responsible for the evolution of new kinds of organisms; 'microevolution' refers to changes in gene frequencies in populations due to migration of individuals between populations (gene flow), drift (random differential reproduction of individuals), mutation (changes in genes), and selection (differential reproduction of individuals due to their differences in ability to cope with problems). Population genetics developed as a subdiscipline within evolutionary biology with a focus on identifying the interaction of those four evolutionary agents and attempting to predict short-term changes in gene frequencies in extant populations brought about by them (Provine 1971). As a result of the emphasis on the study of speciation and population genetics, the study of the profound implications of evolutionary theory—particularly the ability of selection to form adaptations—has, until recently, been relatively unexplored.

As we mentioned in chapter 1, adaptation did not re-emerge as the primary focus of evolutionary theory until the 1960s, largely because of the popular misconception that adaptations function for the good of the group rather than for the individual's reproductive success. The revolution that followed the refocusing of evolutionary thought on the individual level of selection had its greatest influence on the study of social behavior. Social behavior was central to this revolution because, according to modern evolutionary principles, selection at the individual level directly favors only traits that promote the individual's reproductive success, and any group benefit associated with such traits is incidental to the action of individual selection.

Researchers who understood the power of individual-level selection argued that even the altruism observed in the animal world must have somehow served the reproductive interests of individuals in order to have evolved.[1] These researchers became known as *behavioral ecologists*, *sociobiologists*, or (more recently) *evolutionary psychologists*. Their focus on explaining altruism (especially human altruism) as a product of individual selection was at the root of much of the initial conflict between these researchers and those who had not adopted the theoretical advance of individual selection. The individual selectionists' dismissal of group selection and of the notion that selection favored individuals who sacrificed their own reproductive interests for the good of the group, and their alternative view that altruistic acts had actually led to the reproductive success of the altruistic individuals, somehow made the supporters of individual-level selection appear antithetical to many deeply held ideological convictions, including those of Christians, Marxists, and New Age pagans.

Many of the social sciences, especially sociology and cultural anthropology, had been based largely on the assertion that the behavior of individuals functioned for the well-being of organism-like groups such as "cultures" and "societies" (Murdock 1972). Some of the work done in these fields rested explicitly on V. C. Wynne-Edwards's theory of "group selection" (Rappaport 1967; Forman 1967), but more often the connection with that theory was only implicit, the social scientists being unaware that such a theory had been formally proposed and had been rejected by the majority of evolutionists (McCay 1978; Palmer 1994; Palmer et al. 1997). Academics in these social sciences did not greet the bearers of the news about the evolution of altruism with open arms. Indeed, the controversy that arose when researchers finally applied Darwinian individual-level selection to human behavior has been one of the most heated in the history of science (Sahlins 1976; Gould and Lewontin 1979; Kitcher 1985; Rose 1998; for summaries, see Ridley 1993; Wright 1994; Dennett 1995). Although the intensity of the controversy diminished somewhat in the mid 1980s, the debate still has profound effects on the study of human behavior. Wright (1994, pp. 6–7) comments: "People sometimes ask: What ever happened to sociobiology? The answer is that it went underground, where it has been eating away at the foundations of academic orthodoxy." Since, as Wright points out, sociobiology is now re-emerging, primarily under the new label of "evolutionary psychology," there is a renewed need to persuade social scientists to at least reconsider this approach. Thus, we will now identify in detail the other major reasons why so many social scientists seem to suffer from what Daly and Wilson (1988) have termed "biophobia." We will review the misconceptions embraced by social scientists one by one.

The Naturalistic Fallacy

Perhaps the most common misunderstanding of evolutionary theory, and the one most destructive to knowledge, is the naturalistic fallacy: the view that what ought to be is defined by what is, and especially by what is natural (Moore 1903). The flaw in this view seems obvious when one considers such natural phenomena as diseases, floods, and tornadoes. Nonetheless, many of sociobiology's early critics urged its rejection on the unsupportable ground that sociobiological explanations for undesirable traits excused the perpetrators because they were only doing what was natural (Sahlins 1976; Gould and Lewontin 1979).

Even though the naturalistic fallacy has been painstakingly explained in nearly every major work of the past 25 years in which modern evolutionary theory has been applied to human behavior (see, e.g., Alexander 1979, 1987; Symons 1979; Wright 1994), this fallacy continues to be committed by many opponents of the modern evolutionary approach to human nature. For example, Tang-Martinez (1997, p. 117) states that many branches of feminism contend that human sociobiology "serves only to justify and promote the oppression of women by perpetuating the notion that male dominance and female oppression are natural outcomes of human evolutionary history."[2]

One reason the naturalistic fallacy is so prevalent in criticisms of sociobiology is that it is so often committed in the writings of social scientists. Some social scientists have felt free to produce ideological statements about how people ought to behave and then to use the naturalistic fallacy to "justify" their positions on proper and improper human behavior by means of claims (often inaccurate) about the nature of the world. Leslie (1990, p. 896) admits that "most of the influential work in the social sciences is ideological, and most of our criticisms of each other are ideologically grounded." "Our claim to being scientific," Leslie continues, "is one of the main intellectual scandals of the academic world."

The prevalence of the naturalistic fallacy in the social sciences is evident from the frequent use of the same label (e.g., "Marxist anthropology" or "feminist anthropology") to identify both alleged scientific theories about how the world is and ideological positions about how it ought to be. For example, Buss and Malamuth (1996, p. 3) point out that "feminism shares with evolutionary psychology a concern with describing and explaining what exists, but it also carries a social and political agenda." "Hence," they conclude, "feminism is partly concerned with what ought to exist." In such a situation, ideological positions about how the world ought to become are often used to determine the "truthfulness" of statements about how it is.

It is not surprising that some social scientists, having long adhered to the naturalistic fallacy, assume that the statements made by evolutionists about how the world is are intended to imply a position about how the world ought to be. This long history of equating scientific statements and ideological positions in their own work may also explain why some social scientists continue to react to challenges to their statements about the world issued by evolutionarily informed scientists as if they were also challenges to their ideological views.

The prevalence of the naturalistic fallacy can have very specific influences on purportedly scientific statements. For example, the cultural anthropologist Marvin Harris, a leading proponent of the scientific study of human behavior, defines human nature in a manner that is in essential agreement with evolutionary psychology: "As a result of natural selection, our bodies possess a number of specific urges, needs, instincts, limits of tolerance, vulnerabilities, and patterns of growth and decay, which, in sum, roughly define what one means by human nature." (Harris 1989, pp. 126–127) For example, when discussing food, Harris postulates a number of specific evolved "biopsychological components" instead of a general desire to eat. These "components" include numerous mechanisms having to do with how much food is consumed at a given time, precise mechanisms for storing food in the form of fat, and a variety of food preferences (ibid., pp. 142–168). Further, there is a striking similarity between Harris's discussion of children's craving for sweet things and the way a number of evolutionary psychologists have used very similar examples to clarify the basic logic of modern evolutionary theory (Barash 1979; Symons 1979; Wright 1994). However, Harris is actually a leading *opponent* of evolutionary psychology. He vehemently rejects the involvement of specific evolved mechanisms in certain other aspects of human behavior. In contrast to his analysis of the evolved mechanisms involved in eating, Harris proposes that human sexual desires are produced by a mechanism so general that it is shared by males and females. In a chapter titled "Sperm versus Egg?" Harris rejects the Darwinian notion that selection has produced different adaptations in the sexual and romantic desires of men and women. Though he admits that in some societies men and women appear to differ in desire for a variety of partners, Harris dismisses the evidence: ". . . I believe that if they were really free to choose, women would choose as many partners as men choose when they are free to do so" (1989, p. 254). Thus, Harris would have us believe that humans are products of an evolutionary process that designs many precise adaptations for the consumption of food but only a single adaptation for sex—and one so general that it works to overcome the dramatically different obstacles to reproduction faced by males and females.

If Harris realizes that natural selection has provided us with not a general desire to eat but with a complex of specific mechanisms that influence eating behavior, why doesn't he also accept the existence of equally complex and specific mechanisms involved in sex, which might vary between males and females? It may have something to do with the social consequences of these two explanations in a world where many of his colleagues commit the naturalistic fallacy. To propose specific, evolved mechanisms involved in the relatively ideology-free act of eating will arouse no indignation from one's fellow social scientists. However, to propose specific evolved mechanisms *that differ between males and females* is sure to be labeled "sexist" by ideologues who commit the naturalistic fallacy.

The Myth of Genetic Determinism

The naturalistic fallacy is often intertwined with the equally erroneous view that evolutionary explanations are based on the assumption that behavior is genetically determined (meaning rigidly fixed by genes and hence not alterable except by changing those genes). Although the myth of genetic determinism has also been debunked countless times by evolutionists, the psychologist Russell Gray only recently stated that, typically, "evolutionary explanations are [still] taken to imply that our behavior is, in some way, programmed by our genes, and thus the behavior is natural and immutable" (1997, p. 385). Pointing out the absurdity of this situation, the evolutionary biologist John Maynard Smith called genetic determinism "an incorrect idea that is largely irrelevant, because it is not held by anyone, or at least not by any competent evolutionary biologist" (1997, p. 524). "The phrase 'genetic determinism'," Maynard Smith continued, "is one that is usually met in the writings of those who criticize sociobiology, or behavioral ecology."

The chairperson of a university's Women's Studies department recently told us that Donald Symons's book *The Evolution of Human Sexuality*

(1979) was unsuitable as a classroom text because it is based on the premise that behavior is genetically determined. We pointed out that Symons goes into great detail about why such a position is inaccurate (pp. 31–39), and that neither his book nor any other recent evolutionary book we knew of claimed that any behavior was genetically determined. That clearly astonished the chairperson, who evidently had read many claims that genetic determinism was one of the major tenets of the evolutionary approach but who clearly had not read *The Evolution of Human Sexuality.*

Genetic determinism is closely allied to the naturalistic fallacy. According to the myth of genetic determinism, if an author says that an evolved behavior is genetically determined, then the behavior must exist, and therefore we have to accept its inevitability—which amounts to nearly the same thing as saying that the behavior *should* exist. If evolutionists really held this position, their work would be fatally flawed. In fact, however, evolutionists operate without either genetic determinism or the naturalistic fallacy. Most evolutionary works on humans (including ours—see chapter 1) include an extended discussion of the inseparable and equally important influences of genes and environment in the development (ontogeny) and inheritance of all traits of individuals, including cultural or socially learned behaviors.

Failure to Understand Proximate and Ultimate Levels of Explanation

Many social scientists do not understand the distinction between proximate and ultimate explanation—perhaps because their training deals nearly exclusively with proximate explanations (indeed, with a restricted subset of proximate causes), while ultimate questions are dealt with vaguely if at all. This is not surprising; in the absence of an explicitly Darwinian approach, the only alternative ultimate explanations are supernatural explanations or explanations based on some form of implicit group selection (Palmer 1994).

Lack of familiarity with ultimate causation leads many social scientists to mistake evolutionary explanations for proximate ones. This is why the explanation that a behavior exists because it was favored by selection (an ultimate hypothesis) is often mistakenly seen as an alternative to the explanation that learning is involved in the occurrence of the behavior (a proximate hypothesis).

Neglect of ultimate causation has also allowed social scientists to continue to give credence to many proximate explanations despite their implausibility, which becomes instantly apparent once one begins to think of evolutionary implications. That is, there is an obvious incompatibility between our knowledge of how evolution works and many of the specific proximate causes proposed by social scientists. Harris's claim of identical desires for partner variety in males and females is an example. Similarly, the Oedipus complex proposed by Freud would never had been given any credence if anyone had considered the evolutionary fate of a trait that produced such incestuous desires (Thornhill and Thornhill 1987). Because of the reduced viability of offspring produced by mating of close relatives, close inbreeding is selected against. Thus, Freud postulated as fundamental to human nature a trait that simply cannot exist as an evolved human psychological adaptation.

One consequence of the failure to understand the distinction between proximate and ultimate causation is the frequent mistake of assuming that an evolutionary explanation implies a conscious intent to reproduce. Those who make this assumption are evidently not aware that evolutionary theory predicts that individuals will be adaptation executors rather than maximizers of personal reproductive success (Tooby and Cosmides 1992, p. 54). The sticks and carrots of everyday human life are among the events that motivated our ancestors to reproduce successfully, but there was and is no need for individuals to be aware of that result. As Alexander (1979), Dawkins (1986), and other evolutionists have emphasized, we are not evolved to understand that our striving reflects past differences in the reproduction of individuals. Such knowledge can come only from a committed study of evolutionary biology.

Lack of familiarity with ultimate causation also leads many social scientists to think that an adaptation must be an aspect of an organism that increases *current* reproductive success. Hence, social scientists often claim that lack of current reproductive advantage is evidence of a lack of adaptation, or even that it precludes the validity of evolutionary analysis. This has led to many misguided assertions that the inability of a trait to increase reproductive success in a current environment is evidence against the trait's being an adaptation. For example, Tang-Martinez (1997, p. 140) emphatically rejects Wright's (1994) argument that the marriage of Aristotle

Onassis and Jacqueline Kennedy was an example of a wealthy, high-status male attracting a younger woman on the grounds that "they had no children" and "it is unlikely that Onassis's wealth contributed anything to Jackie's reproductive success."[3]

Similarly, some critics have told us that an evolutionary analysis of rape will fail because rape hasn't been shown to yield much reproductive success in the United States or in some other modern environmental setting. We were dumbfounded to hear this argument when we first tried to publish a paper on human rape in its evolutionary context, especially from an editor of a major biological journal (*Animal Behaviour*). Apparently, the editor didn't like our paper because it contained no data on the current reproductive success of rape, which the editor erroneously thought to be crucial to the hypothesis that rape was associated with enhanced reproductive success during human evolutionary history.

As was implied by our treatment of rape-related pregnancy in the preceding chapter (which was focused entirely on whether rape results in pregnancy despite any evolved female defenses to control fertilization), the coevolutionary battle of the sexes is ongoing, and which sex is ahead at any time is largely unpredictable (Parker 1979; Clutton-Brock and Parker 1995). Studies of rape-related pregnancy indicate that human rapists are still in the evolutionary game—that women's cryptic sire-choice adaptations against rape don't entirely prevent conception or birth of offspring after rape.

Failure to fully understand ultimate causation also leads to the misguided position that evolutionists assume every aspect of an organism to be an adaptation. For example, Sork (1997, p. 110) states that a "major weakness of sociobiology and evolutionary psychology is that all behaviors are seen as adaptations."[4] In fact, as we have noted, George Williams made distinguishing adaptations and incidental effects a central goal of the study of adaptations in 1966.

A related and equally incorrect claim is that evolutionary explanations are tautological. That would be true if such explanations assumed that the mere existence of any trait was sufficient proof that the trait had been formed by direct natural or sexual selection, but such is not the case. Similarly, it is not true that evolutionary explanations are "just-so stories" (Gould and Lewontin 1979; Avise 1998)—that is, assertions about the evolutionary function of a trait that are unskeptically accepted by evolutionists. The strength of evolutionary theory lies in the fact that, in addition to generating testable alternatives (ultimate hypotheses about the function of an adaptation), it provides criteria for determining whether a given aspect of an organism even is an adaptation (as opposed to a byproduct or a side effect of other adaptations, or a product of drift), or whether it is explicable by mutation-selection balance.

Another reason why many social scientists cannot seem to understand that there are ultimate causes to human behavior may stem from evolved human psychological intuitions (Pinker 1997). Because of past natural selection for ability to predict what other people will do, humans are expert at inferring certain proximate motivations and emotional responses from facial expressions and body movements (Humphrey 1980; Pinker 1997). The fact that humans are good at these tasks may explain why some individuals genuinely feel that they have valid opinions about all explanations of human behaviors and psychological states. If you inform such a person of an evolutionary theory of rape, child maltreatment, or human moral behavior, he or she is likely to pass judgment on it immediately—something that few people would do upon encountering the supersymmetry theory of physics. This difference cannot be attributed to a difference in the complexity and diversity of the phenomena or the theories; evolutionary theory deals with the most complex and diverse set of phenomena in all of science. (See, e.g., Dawkins 1986.) Evolved psychological intuitions about behavioral causation can mislead individuals into believing that they know as much as experts do about proximate human motivation.

The impulsiveness with which many individuals judge a scientific idea about human behavior or psychology may arise from their motivation to display their belief system or ideology as moral. Furthermore, articulated affiliation with an ideological group and display of concern for others appear to be human universals.

The Perceived Threat to Ideology

Much of the opposition to evolutionary theory has been based on ideological grounds. Some Marxists have willfully misinterpreted evolutionary explanations of human behavior as assertions of support for the political status quo and as a rationale for opposition to change. Some individuals who see ending the oppression of women as a political ideology have misinterpreted evolutionary explanations of differences between men and women as claims that equality for women is bad because it is not natural, or that it is impossible because male and female differences are genetically determined. Ideology also appears to be behind rejections of the evolutionary explanation on the ground that it is reductionist and thus bad science.

Although 'reductionism' is seen as a vulgar word in some areas of the academy, reductionism is actually an essential aspect of all science. All scientific hypotheses attempt to elucidate nature by simplifying its complexity and its diversity into empirically manageable parts (Williams 1985; Wilson 1998). That such a procedure generates discoveries is demonstrated by the vast amount of knowledge produced by scientific disciplines. Negativity toward reductionism seems most likely to arise when certain ideas are perceived as threatening certain ideological positions (Lewontin et al. 1984; Rose 1998).

Another form of ideology that has recently gone to war with evolutionary theory goes by the amorphous name *post-modernism*. This position, founded on a liberal ideology that proposes no censorship or even no critical evaluation of ideas, sees itself as a response to the totalitarianism inherent in typical ideologies, especially those of Western societies (Murphey 1992; Wilson 1998). To a post-modernist, scientific findings about the world, such as those generated by evolutionary theory, have no more accuracy or validity than literature, theology, or creation myths. Such a position may be useful in persuading people to adopt certain ideologies, but it prevents the accumulation of knowledge.

Threats to Status and to Altruistic Reputation

Because the evolutionary approach threatens the theories and approaches that have traditionally been used to study human behavior, it poses a serious threat to the status of those who have achieved success in their fields using non-evolutionary approaches. This intensifies the resistance to evolutionary theory of many social scientists, for whom accepting the evolutionary approach would amount to admitting that their previous approach was not valid. Moreover, such an admission would move them from the expert level to the level of a beginner.

Yet another reason for resistance to evolutionary theory appears to be that humans have evolved to present themselves as altruists (Alexander 1987; Nowak and Sigmund 1998). This presentation involves what Alexander (1987) calls "indirect reciprocity," in which the displayer obtains a reputation or a social image as a potential cooperator. Here the word 'indirect' pertains to the fact that the returns to the altruistic individual come from onlookers rather than from the individual helped, whereas in direct reciprocity the returns come directly from the individual helped. Altruistic display, in which one attempts to portray oneself as concerned about anonymous others and often about the human species in general, is crucial to an individual's development of a reputation as a social ally. Announcing that one accepts the modern evolutionary premise that altruistic acts were selected only when they were actually reproductively self-serving in the past may interfere with this social strategy. We hypothesize that people's desire to present themselves as moral and benevolent, and thus as useful to others in reciprocal cooperation, is a fundamental reason why controversy surrounds biology's finding that evolved altruism is reproductively advantageous to altruists. Also, this is likely a proximate reason why many social scientists and early biologists embraced Wynne-Edwards's theory of group selection, with its assertion that humans have evolved to do things for the good of the group, and why some biologists still retain this notion (Wilson and Sober 1994; Sober and Wilson 1998), even in the presence of voluminous data against the effectiveness of group selection in the evolution of an adaptation and of a theory that explains the data.[5]

These threats to status and reputation may help explain the paradox that perhaps the best-known evolutionist of our time is a leading critic of the use of evolutionary principles to explain human behavior. Stephen Jay Gould began to attack evolutionary explanations of human behavior after the publication of biologist Edward O. Wilson's book *Sociobiology* (1975).[6] Even today, Gould "carries on unfazed, repeating to a large general audience the same arguments that his colleagues in evolutionary biology . . . examined, refuted and dismissed long ago" (Alcock 1998, p. 3). What makes this so paradoxical is that Gould clearly understands the difference between proximate and ultimate explanations, that evolutionary explanations don't imply genetic determinism, and that the naturalistic fallacy is a fallacy. For example, he states that "every scientist, indeed every intelligent person, knows that human social behavior is a complex and indivisible mix of biological [read 'genetic'] and social influences" (Gould 1987, p. 113), and that "nature has no automatically transferable wisdom to serve as the basis of human morality" (ibid., p. 225). Hence, Gould must "know what he is doing" (Alcock 1998, p. 6) when he caricatures the sociobiological explanation of male-female differences as being based on genetic determinism and imbued with the naturalistic fallacy: "The sociopolitical line of the pop [sociobiological] argument now leaps from the page: males are aggressive, assertive, promiscuous, overbearing; females are coy, discriminating, loyal, caring—and these differences are adaptive, Darwinian, genetic, proper, good, inevitable, unchangeable. . . ." (Gould 1987, p. 36)

Gould's misrepresentation of evolutionary biology has been the object of many critiques by evolutionary biologists.[7] In particular, Dennett (1995) has detailed the unscrupulous tactics Gould has used in his attempts to discredit the evolutionary explanations that Gould apparently thinks threaten certain political ideologies. Dawkins (1986), Dennett (1995), Pinker (1997), and Waage and Gowaty (1997) have all pointed out that the major theoretical breakthroughs Gould has claimed to have made are actually all established parts of the evolutionary writings he is attempting to discredit.[8] Others critics have been even harsher. Wright's (1990) review of Gould's popular book *Wonderful Life* was titled "The intelligence test"; according to Wright, Gould failed. Of Gould's work as a whole, Maynard Smith (1995, p. 46) has written the following: ". . . the evolutionary biologists with whom I have discussed his work tend to see him as a man whose ideas are so confused as to be hardly worth bothering with. . . ."

Gould's 1979 paper (with Richard Lewontin) "The spandrels of San Marco and the Panglossian paradigm" has become a classic of rhetorical argumentation.[9] Gould's literary talent has aided his construction of a caricature of "evolutionary biology" that seems plausible to many people who are not immersed in the facts and theory of evolution or who find his critiques ideologically supportive. Although a number of evolutionists have criticized Gould and Lewontin's infamous paper on scientific grounds, biologist David Queller's (1995) critique is the most complete and definitive exposition of the paper's ideological basis and its misrepresentation of scientific knowledge.

Our main goal in describing the "Gould phenomenon" is neither to criticize Gould's science nor to praise his fiction; it is to get future potential critics of evolutionary explanations of human behavior to question how ideology and threats to status and reputation may affect their thinking before they reject what evolutionary theory has to offer. We also hope to influence future critics of the evolutionary approach to consider the extent to which their opposition is based directly on Gould's highly suspect pronouncements. As Alcock (1998, p. 17) has noted, Gould's "constant and combative repetition of anti-adaptationist arguments has certainly influenced the general public's understanding of evolutionary science as well as encouraging social scientists to think that they can ignore evolutionary biology."

Gould's approach has appealed to some biologists who, on ideological grounds, take the position that evolution applies to all life except human behavior and psychology. This bizarre position has a distinguished history. The biologist A. R. Wallace (a contemporary of Charles Darwin who independently discovered natural selection's role in evolution) held it; so does Pope John Paul II, whose recent "cautious endorsement of evolution as a factually supported theory," Field (1998, p. 296) notes, "restricted its explanatory powers to the physical realm: 'theories of evolution which . . . consider the mind as emerging from the forces of living matter . . . are incompatible with the truth. . . .'"

The pre-Darwinian position that human behavior is exempt from evolutionary analysis—held by many biologists—is one reason why many social scientists do not understand the power of Darwinism for explaining, and making new discoveries about, human activity. One of us (Palmer) works in an anthropology department, the other in a biology department. In view of the fact that the subdisciplines of anthropology that treat human behavior and psychology have always been divorced from evolutionary biology (Brown 1991), it is not surprising that Palmer is viewed as a bit unusual by some of his colleagues for being interested in the evolution of human behavior. But Thornhill also encounters criticism from some members of his biology department, and even from some biologists who study *non*-human behavior and psychology from an explicitly evolutionary perspective. These critics sometimes say that the study of humans should not be part of a biology department, or that papers on human behavior should not be included in a scientific conference on the behavior of non-human animals.

Opposition to Evolutionary Explanations of Rape

All the aforementioned misunderstandings of evolutionary theory were evident in the critiques of the very first evolutionary analyses of rape to be published: Shields and Shields 1983, Thornhill 1980, Thornhill and Thornhill 1983, and Thiessen 1983/1986. (See especially Baron 1985; Dusek 1984; Fausto-Sterling 1985; Kitcher 1985; Sunday and Tobach 1985.) Underlying all the opposition to those papers was the naturalistic fallacy. Opponents of the evolutionary approach asserted that it legitimized rape.

Criticizing an earlier paper (Thornhill 1980), Gowaty (1982) and Harding (1985) accurately observed that the inclusion of evolutionary function in the definition of 'rape' used in that paper was not consistent with how the word is generally used in human affairs. However, the inclusion of an evolutionary angle in a definition provides no rational reason for thinking that the definition's author had a hidden desire to justify rape by calling it natural or evolved. Indeed, evolutionary biologists often use an evolutionary explanation when first defining a behavior (e.g., nepotism, reciprocity, selfishness, mating, rough-and-tumble play). Further, the misuse of the word 'definition' to refer to what is actually a possible evolutionary function is typically obvious to those not opposed to the entire study of rape for ideological reasons. Consider that Smuts and Smuts (1993, p. 2) recently "defined" sexual coercion as "the use by a male of force, or threat of force, that functions to increase the chances that a female will mate with him at a time when she is likely to be fertile, and to decrease the chances that she will mate with other males, at some cost to the female."[10] Although it might be advantageous to a researcher to acknowledge explicitly that an act would still be considered sexual coercion whether or not it occurred when a female was likely to be fertile, it is quite obvious that this part of the statement simply adds an evolutionary hypothesis, not a moralistic justification, to a definition of "sexual coercion."

The naturalistic fallacy is even clearer in the obviously invalid objections to claims that rape occurs in non-humans. Many social scientists, when criticizing the evolutionary approach to human rape, bring up the problem of "extrapolating from animal and insect behavior" (Polaschek et al. 1997).[11] Biologists are regularly derogated for being so naive as to think that findings on non-human animals, especially insects, have any bearing on understanding *Homo sapiens*. These critics claim that if other animals rape then human rape is natural and thus justifiable. For example, Baron (1985, p. 273) states that the "claim that rape occurs widely among plants and animals" is "likely to trivialize the meaning of rape and give it a veneer of justifiability."[12] This is the naturalistic fallacy in its purist form.

Although its illogicality is apparent to anyone who realizes that the naturalistic fallacy is a fallacy, the fear of "justifying" rape has led many critics of the evolutionary explanation to try to restrict the definition of rape to the human species. Many critics (Burns et al. 1980; Estep and Bruce 1981; Gowaty 1982; Hilton 1982; McKinney and Stolen 1982; Dusek 1984; Baron 1985; Blackman 1985; Harding 1985; Kitcher 1985; Sunday 1985; Tang-Martinez 1996) have protested that, owing to its uniquely human connotations, the word 'rape' should not be used in reference to non-humans under any circumstances. In reality, however, though many people may initially think of human examples upon hearing the word, that is no reason to restrict its application to humans. Many people probably also think of human examples upon hearing the word 'sex', but that word is routinely applied to other species without misunderstanding. "Forced copulation" (which some have claimed is the proper term for rape in non-human species) is inadequate as a definition of rape because many rapes involve only the *threat* of force, and asserting that the same behavior be called by one name in reference to non-human species and by another in reference to humans only confuses matters unnecessarily. Furthermore, asserting that rape is by definition unique to humans excludes the behavior of non-human animals as a potential source of information about the causes of human rape. Indeed, it denies the importance of comparative analysis, which is a fundamental tool in biology for understanding causation (Crawford and Galdikas 1986; Palmer 1989b).[13]

Many of the same critics cited above also attacked the hypothesis that rape might be an evolved adaptation in certain species. Evidently, they assumed that, if rape was favored by "natural" selection, it must be "natural" and hence good or at least excusable. Their arguments demonstrated a lack of understanding of what 'adaptation' means in evolutionary biology. For example, several critics argued that the relatively low rate of rape-related pregnancy was evidence that rape could not be an evolved adaptation in humans (Dusek 1984; Harding 1985; Sunday 1985). Not only does that argument assume that *current* benefits to reproduction define an adaptation; it also fails to consider that a relatively low rate of reproduction does not necessarily falsify an explanation of rape as an adaptation any more than a low rate of reproduction demonstrates that male ejaculation is not an adaptation (Palmer 1991). Moreover, even traits that confer a seemingly trivial net reproductive benefit (say, 1 percent) relative to alternative traits increase in frequency very rapidly as a result of evolution by selection (Bell 1997).

Critics also focused on the role of sexual motivation in the occurrence of rape. Here the naturalistic fallacy ("If rape is sexually motivated, it is natural and must be good") and genetic determinism ("If rape is a 'biological' act of sex, it must be inevitable") combined with a failure to understand the difference between proximate and ultimate explanations. For example, Dusek (1984, p. 10) stated that "sociobiologists claim that rape is primarily a 'strategy' for reproduction and, further, that it is an erotic rather than a violent act, thereby nullifying educative attempts of the anti-rape movement to the contrary." Whether or not sexual desire plays a role in the motivation (proximate cause) of rape is indeed a crucial question with far-reaching practical implications, but it is a different question than whether rape reflects adaptation for rape itself or whether it is a by-product of other adaptations.

The following quote from a typical social science book on rape shows, all at once, evidence of the naturalistic fallacy, the myth of genetic determinism, the failure to distinguish ultimate from proximate explanations, the false dichotomy between learned and genetically affected behaviors, and the assumption that an ultimate explanation implies current maximization of fitness:

> One of these theories is the sociobiologists' viewpoint on rape. Though this theory has the most repugnant social implications, it has won many adherents in the Western world over the past decade, particularly among those who hold traditional pa-

> triarchal views of human society. . . . Sociobiologists offer what amounts to an evolutionary justification of rape. According to this perspective, rape is simply one way for males cut off from socially acceptable access to female sexual partners to ensure that their genetic endowment is passed on to another generation. To suggest, however, that rapists are driven by some genetic force beyond their control is untenable. . . . That relatively few men rape (though almost certainly more than are ever reported and caught), and that these men rape only under conditions where they are likely to get away with it, indicates that this behavior is very much learned, not genetically inherited. (Marshall and Barrett 1990, pp. 105–106)

Summary

As Wilson et al. (1997, p. 433) note, "Darwinian selection is the only known source of the functional complexity of living things, and biologists have no reason to suspect that there are any others." Social scientists should no longer refuse to acknowledge this, nor should they refuse to acknowledge that our understanding of evolution—especially the evolution of adaptations associated with human social life (sex, nepotism, reciprocity)—has increased dramatically in the last 35 years as a result of the intellectual revolution in biology caused by the realization that selection was most potent at the level of the individual.

Rejections of evolutionary theory based on the naturalistic fallacy, and hence on ideological grounds, are not tenable. The naturalistic fallacy has been described and discredited so many times that those who continue to evince it in their critiques of evolutionary explanations should be dismissed on the basis of lack of scholarship alone.

The biophobia that has led to the rejection of Darwinian analyses of human behavior is an intellectual disaster not only because it has discouraged the accumulation of knowledge but also because of what it has allowed to pass for knowledge. Most of what is scientifically inaccurate and counterproductive about how the social sciences and academic feminism approach the study of rape stems directly from the aversion to modern theoretical biology in those fields.

6

The Social Science Explanation of Rape

Zuleyma Tang-Martinez's phrase "a feminist psychosocial analysis" (1997, p. 122) accurately describes what has become the dominant explanation of rape in the social sciences over the past 25 years. This explanation developed after certain feminist assertions were added to the "learning theory" that has been the bedrock of social science for much of the last 100 years. Because the phrase "feminist psychosocial analysis" is a bit awkward, we will refer to it as "the social science explanation."

The social scientists we mean are those whose "research" has been guided more by ideology-driven social arguments than by science. For a definition of feminism, we rely on Gowaty (1992, p. 218): "a movement to end sexist oppression."

We have been told that some of the positions we are about to criticize have been abandoned by social scientists studying rape. We are not convinced. Not only does the recent literature on rape repeat these positions; assertions reflecting these positions often continue to be made by the same people who claim that the arguments have been abandoned (Palmer et al. 1999). Hence, we feel our use of the label "the social science explanation" is quite justified.

Learning Theory

The social science approach to rape is based on learning theorists' assertion that culture is a non-biological entity and that it causes the behavior and the desires of men and women through a powerful process known as "learning."[1] Hence, rape occurs only when men learn to rape. One reason why so much of social science pays little attention to scientific standards is that this "learning theory" is almost metaphysical, so that making the implicit evolutionary assumptions of the learning theory of rape explicit is especially challenging.

In its extreme form, learning theory holds the view that, as the cultural anthropologist Clifford Geertz has put it, "our ideas, our values, our acts, even our emotions are, like our nervous system itself, cultural products" (quoted in Ehrenreich and McIntosh 1997). Hence, although learning theorists claim to be scientists and typically claim to accept evolution, learning theory has many similarities to a religion. Indeed, it has been labeled "secular creationism" (Ehrenreich and McIntosh 1997), because it proposes a supernatural (or at least a "superorganic") "creator" of all human behavior: culture. Another similarity to religion is that this "creator" is often alleged to work in a mysterious way ("learning") through arbitrary environmental experiences to make human brains and thus human behavior.[2]

"Rape Is Not Sex"

To the general framework of learning theory many feminist social scientists added the assertion that "sexual coercion is motivated by power, not lust" (Stock 1991, p. 61). This addition made male dominance a larger target of feminist opposition. This view was first put forth by Millett (1971), Griffin (1971), and Greer (1970). When popularized by Brownmiller (1975), it quickly became the central tenet in social science explanations of rape. As Warner (1980, p. 94) notes, "it is now generally accepted by criminologists, psychologists, and other professionals working with rapists and rape victims that rape is not primarily a sex crime, it is a crime of violence." Further, the idea of rape as "a political act that indicated nothing about male sexuality" (Symons 1979, p. 104) became a "focal point of feminist theory" (Sanders 1980, p. 22). Indeed, by the mid 1980s rape had become the "master symbol of women's oppression" (Schwendinger and Schwendinger 1985, p. 93). This intertwining of explanations of rape and political ideology has caused the naturalistic fallacy to play a truly impressive role in the social science study of rape.

In combination, learning theory and the feminist assertion that rape is motivated by a desire for control and dominance produced the view that rape is caused by supposedly patriarchal cultures where males are taught to dominate, and hence rape, women. For example, Stock (1991, p. 61) states that the feminist–learning theory approach[3] "defines sexual coercion as power motivated, upholding a system of male dominance." Similarly, Sorenson and White (1992, p. 3) state that "patriarchy shapes attitudes and beliefs, women's roles, men's roles, and their relationship to each other, ultimately determining all forms of violence against women." Hence, rape supposedly occurs in cultures like that of the United States for the following reason:

> Male sex-role socialization [requires] that males separate their sexual responsiveness from their needs for love, respect, and affection. . . . Males are trained from childhood to separate sexual desire from caring, respecting, liking or loving. One of the consequences of this training is that many men regard women as sexual objects, rather than as full human beings. . . . [Male sex-role socialization] predisposes men to rape. Even if women were physically stronger than men, it is doubtful that there would be many instances of female rapes of males: Female sexual socialization encourages females to integrate sex, affection, and love, and to be sensitive to what their partners want. (Russell 1984, pp. 119–120)

It is difficult to overestimate the power the "not sex" theory of rape continues to have. Murphey (1992, p. 18) offers a typical example from the popular press: "Joan Beck, a nationally published columnist based with the *Chicago Tribune*, was able to say in April 1991 that 'if there is still any lingering misconception that rape is a crime of sexual passion, it's important to drive a stake through the heart of that idea as quickly as possible. . . .'" Jones (1990, pp. 64–65) explicitly praises Brownmiller's 1975 book for having taught feminists that "sexual and physical violence against women is not 'sexual' at all but simply violent."[4] Sanday (1990, p. 10) states that during rape "the sexual act is not concerned with sexual gratification but with the deployment of the penis as a concrete symbol of masculine social power." Donat and D'Emilio (1992, p. 15) write that among feminists in the 1960s "rape was recognized as an act of violence, not of sex." White and Farmer (1992, p. 47) state that "feminist assumptions . . . generally de-emphasize the potential contribution that biologically driven sexual motives may play in the commitment of sexual assault." Morris (1987, p. 128) writes that "most feminist writers . . . see rape as a violent act [and] argue that the use of force or physical coercion is the central feature of rape."

As an indication of the strength and the pervasiveness of these assumptions among social scientists,[5] consider the following quotation:

> We cannot overestimate the influence of feminist theorists such as Brownmiller upon the thinking of current researchers. Many investigators, while not necessarily testing the assumption in their studies, presume that rape is a manifestation of male dominance over and control of women." (Sorenson and White 1992, pp. 3–4)

That "Brownmiller's book established decisively that rape is a crime of violence rather than passion" (Buchwald et al. 1993, p. 1) is the starting point for most feminist studies of rape, since that book "has been recognized as the cornerstone of feminist scholarship on rape" (Ward 1995, p. 19). In 1992, Susan Brownmiller herself endorsed the "feminist" view she had popularized in the 1970s: "The central insight of the feminist theory of rape identifies the act as a crime of violence committed against women as a demonstration of male domination and power." (Brownmiller and Mehrhof 1992, p. 382) Davies (1997, p. 133) stated that it is a contention of feminists "that rape is an act of power, not sex." Polaschek et al. (1997, p. 128) argue that feminist theories view rape as primarily motivated by male dominance, and furthermore that evolutionary theories proposing "sexual motivation for rape, with associated aggressive and dominating behavior viewed as tactics rather than goals" are "in stark contrast to feminist . . . and broader social learning positions."

That the "not sex" explanation remains popular among feminists, and that it has dominated feminist writings on rape, is even admitted by some individuals who point out that not all feminists have supported it. For example, Muehlenhard et al. (1996, p. 129) admit that "in general, . . . feminist theorists have emphasized the goals of dominating and controlling rape victims and women in general."

It is also not clear that even the relatively few feminists who appear to disagree with the "not sex" explanation really do so. A minority of feminists, many known as "radical feminists," *appear* to have disagreed with the assertion that rapists are not sexually motivated. (See e.g. MacKinnon 1983, 1987, 1990, 1993; Dworkin 1989, 1990.) However, these social constructionists assert that "male power creates the reality of the world" (MacKinnon 1989, p. 125) and that "sex is a social construct of sexism" (ibid., p. 140). That is, "sex is constructed . . . specifically to be male dominance" (Dworkin 1990, p. 138). In view of this particular social construction, "sexuality . . . is a form of power" (MacKinnon 1989, p. 113), and "violence is sex" (ibid., p. 134).[6] That is, these writers assert that rape is a sexual act, but only "in a culture where sexuality is itself a form of power" (Caputi 1993, p. 7). How literally is the assertion that sex, violence, and power are all the same thing taken by these feminists? Very. For example, MacKinnon (1989, p. 178) states that "a feminist analysis would suggest that assault by a man's fist is not so different from assault by a penis, not because both are violent but because both are sexual." Hence, far from opposing the view that violence and power are the goals of rapists, these authors are actually arguing that power and violence not only are the goals of males engaged in rape; they are also the goals of males engaged in other forms of sex. Instead of really arguing that Brownmiller was wrong about rape, the position of these radical feminists "in effect extends Brownmiller's definition of rape to the sex act itself" (Podhoretz 1991, p. 31). Although the statements of these writers appear to differ from the "not sex" explanation when they are taken out of the context of the social constructionist argument, within that context they actually share the fundamental assumption that rapists are motivated by desire for violence and power—which, according to their theory, just happen to be the same thing as sex.

When some feminists appear to be challenging the dominant feminist position by arguing that rape is about sex, they actually mean that "it is *social* sex, not biological sex, that rape is about" (Bell 1991, p. 88). Further, "social sex" is the motivation for rape only when sex is socially constructed to be the same thing as power and violence. For example, Scully and Marolla (1995, p. 66) state that "from the rapists' point of view rape is in part sexually motivated," but they emphasize that this is only because "they have learned that in this culture sexual violence is rewarding" (ibid., p. 71). Jackson (1995) states that rape is sexual, but only because our culture has given us "sexual scripts" that happen to equate sex with power and aggression. Hence, according to Jackson's theory, rape is impossible when people are given "sexual scripts" in which sexuality is "not bound up with power and aggression." Jackson attempted to validate this position by citing Margaret Mead's (1935) description of "the Mountain Arapesh of New Guinea" as "the most famous example of a society where rape is unknown"—a society in which there is "no element in [the] sexual scripts which could create the possibility of rape" (ibid., p. 27). In reality, however, the Arapesh are quite familiar with rape—see below.

The preceding review of the literature supports Malamuth's (1996, p. 270) conclusion that "although there have been some writers who have emphasized somewhat different perspectives (e.g., MacKinnon, 1989), most feminist writers have defined rape and other forms of sexual coercion as motivated not by sexuality, but by a desire to assert power over women."

Flaws in the Social Science Explanation of Rape

The social science explanation of rape has five major errors:

- The assumptions it makes about human nature are not compatible with current knowledge about evolution.
- Its assertion that rape is not sexually motivated is based on arguments that cannot withstand skeptical analysis.
- Its predictions are not consistent with the cross-cultural data on human rape.
- It does not account for the occurrence of rape in other species.
- It rests on several assertions that belong more to metaphysics than to science.

After describing these errors, we will examine how they are reflected in the empirical data produced by the social science theory of rape. We will then describe how ideological concerns have maintained the popularity of the social science explanation of rape despite its failure as a scientific explanation.

Incompatibilities with Evolutionary Theory

Any explanation of human behavior makes an implicit assertion about human nature. This is because, as the psychologists Martin Daly and Margo Wilson (1996, p. 23) point out, all "sociological, economic, and political hypotheses are necessarily built on implicit psychological hypotheses about how individual human actors perceive and are affected by social, economic, and political variables." Since nearly all social scientists claim to accept evolution, all their explanations of human behavior are implicitly evolutionary. Only if they make their evolutionary assumptions explicit, however, can these theories be properly evaluated in light of the modern understanding of evolution. We will now attempt to make the evolutionary assumptions of the social science theory of rape explicit.

The most fundamental premise of the social science theory of rape—that an individual's psychology is "determined" solely or mostly as a result of socialization—implies something close to the classic *tabula rasa* view of human nature. Based on the false assumption that aspects of living organisms can be divided into biological and non-biological categories, this view holds that human "biological" nature consists of a few basic needs (sex, love, respect, affection), but that these can be combined or separated in any way that the non-biological entity of culture dictates. Indeed, these desires and emotions are sometimes asserted to be present only when a culture dictates their existence. Hence, violent behavior is something that exists only when it is taught, and individuals will find sexually attractive only those beings and other objects in the environment that they are told to find sexually attractive. On the further assumption that there are no differences between the "biological" natures of human males and human females, males and females respond to the same cultural instructions in the same manner.

These propositions are entirely at odds with current knowledge about evolution because they fail to acknowledge that selection has shaped the psychology of human learning and decision making. If the referent of "a culture" is restricted to what can be identified by the senses, a culture is seen to consist of no more than a number of individuals interacting in certain ways. If the psychology and the learning capacities of these individuals are seen as products of selection at the individual level, then a culture is a conglomeration of individuals, each designed to engage successfully in social competition with other individuals (Alexander 1979; Cronk 1995; Flinn 1997). Individuals may form alliances and cooperate, but only when such cooperation is a successful tactic in their competition.

Once the basic premise that human psychology and culture are products of selection at the individual level is understood, the implausibility of the social science view of human nature is revealed simply by asking these questions: What would have been the evolutionary fate of individuals in ancestral populations who possessed the nature implied by the social science explanation? How would they have fared in reproductive competition with individuals who had a more specialized set of psychological adaptations? In particular, what would be the evolutionary fate of individuals who engaged in the high-cost behavior of violence only when they were told to do so by others (including individuals who were their reproductive competitors)? Because violent behavior has very high costs, this would have given a tremendous advantage to the competitors, who could simply instruct susceptible rivals to engage in violence when the potential benefits of such competition were low and to forgo violence when the potential benefits were high. Males who engaged in violence with no benefits simply because they were taught to do so must be no one's evolutionary ancestors, because they soon would have been outreproduced by males with specialized psychological mechanisms predisposing them to engage in aggression only when the benefits outweighed the costs. Indeed, there is much evidence of the finely tuned design of violent behavior in terms of the costs and benefits of aggression as a solution to very specific problems for the aggressor. Evolutionary biologists have studied aggression intensively across a wide variety of animal species, especially in the last 25 years. A rich empirical base supports the evolutionary view that aggression has evolved as a result of selection, and that therefore aggression is conditionally patterned in relation to predictable ecological factors that affect its benefits and its costs. (See, e.g., Elwood et al. 1998.[7])

Reproductive failure also would befall an individual who could be instructed to form cooperative relationships with others who intended to exploit his trust and love. Altruism and cooperation can increase reproductive success only when they are directed toward genetic relatives or toward reliable reciprocators (Trivers 1971). The individual whose psychological systems predisposed him or her to exhibit helpfulness in arbitrary ways, as directed by his "culture," would have been most unlikely to outreproduce competitors. Hence, the individual proposed by social science theory to have such flexible emotions would have become no one's ancestor.

An equally unsuccessful fate would await individuals who were sexually aroused only by individuals they were instructed to desire. Sexual attraction and arousal have many non-arbitrary features, including species, sex, age, and health. (See chapter 2 above.) Males or females in human evolutionary history who mated randomly with regard to any of these characteristics of potential mates are also no one's evolutionary ancestors. Males or females in human evolutionary history who mated only with the mates they were told to mate with by their evolutionary competitors quickly eliminated from the gene pool the genetic basis for the development of this kind of learning "ability." For example, competitors could quickly reduce the reproductive success of such individuals to zero by telling them not to be sexually attracted to desirable members of the opposite sex.

That males and females not only have very flexible general sexual adaptations but have the *same* general sexual adaptations is even less plausible. Males and females in human evolutionary history were presented with very different problems of selecting and competing for mates, and thus women and men have differently designed sexual psychologies. The kind of males and females proposed by social science theorists would have been quickly outreproduced by more specialized males and females whose psychological mechanisms inclined them to behave in ways that solved the sexual challenges facing their specific sex more efficiently.

Rape and Sexual Motivation

As the literature cited above demonstrates, the social science theory of rape rests on the assumption that a non-sexual motivation (such as a desire for power, control, domination, and/or violence) is both necessary and sufficient for a rape to occur. Aside from ignoring evolution and the ultimate level of explanation, this assumption can be accepted only if one accepts a bizarre definition of 'sex', suspends logic in the evaluation of supporting arguments, and abandons all skepticism in evaluating evidence. As the literature cited above, in Palmer 1988a, and in Palmer et al. 1999 demonstrates, many social scientists still imply that sexual desire is not sufficient or even necessary as a motivation for rape. This position, which remains at the heart of the social science explanation of rape, is routinely used to make pronouncements on what individuals ought to do to prevent rape.

There is no question that multiple motivations may be involved in any human behavior. An individual rapist *may* be motivated by a desire for revenge against a particular woman who turned down his earlier sexual advances, by a desire to humiliate or inflict pain on a particular woman or on women in general out of hatred for his own mother, by a desire to impress other males by losing his virginity, or by any of a countless number of other possible motivations. But have social scientists really demonstrated that any rapist is not at least partially motivated by sexual desire? Indeed, could any rape really take place without any sexual motivation on behalf of the rapist? Isn't sexual arousal of the rapist the one common factor in all rapes, including date rapes, pedophilic rapes, rapes of women under anesthesia, and rapes committed by soldiers during war? Further, would a rapist have to have any of the possible non-sexual motivations in order to commit a rape? Isn't it possible for a male's sole motivation for committing a rape to be a desire for sexual gratification?

One reason these seemingly obvious points have been obscured is that social scientists typically present the issue in terms of whether rape is "an act of" sex, "an act of" violence, or both. Perhaps by intention, use of the phrase "an act of" blurs the difference between the goals that provide the *motivation* for rape and the *tactics* used to accomplish those goals. Rape is obviously not the same *act* as consensual copulation, because by definition rape implies the use of certain distinct tactics (e.g., force or the threat of force). But that doesn't mean that the motivation of the male necessarily differs.

The importance of distinguishing between the goals that motivate a behavior and the tactics used to accomplish those goals becomes clear when one considers prostitution. The *act* of prostitution includes both a person giving money to another person and a sexual act. Does this mean that a man who goes to a female prostitute is motivated by a desire to give money to a woman? Does it even mean that the man is motivated by both a desire for sex and a desire to give his money to a woman? A man might have numerous motivations for going to a prostitute, but isn't it possible that the man lacks any desire to give his money to the woman? Isn't it indeed likely that the man gives his money to the woman only as a tactic to gain the desired goal of sex, which is the sole motivation of his behavior? Further, isn't it possible that the man would much prefer to have sex with the woman without having to give her money? If the same "logic" that has been used in the social science explanation of rape were to be applied to prostitution, people would be asserting that going to a prostitute is an "act of altruism, not sex," or at least that it is "an act of both altruism and sex."

A Critique of the Arguments

We now offer a critique of the arguments that are most often used to support the claim that rapists are not sexually motivated.

Argument 1 *"When they say sex or sexual, these social scientists and feminists [who argue that rape is not sexually motivated] mean the motivation, moods, or drives associated with honest courtship and pair bonding. In such situations, males report feelings of tenderness, affection, joy and so on. . . . It is this sort of pleasurable motivation that the socioculturists (and feminists) denote as sexuality. . . ." (Shields and Shields 1983, p. 122)*

The sociocultural definition of 'sex' is inaccurately and unnecessarily restricted. In view of the more common usage of the word 'sex', it is, according to Hagen (1979, pp. 158–159), "abundantly self evident . . . that a large percentage of males have no difficulty in divorcing sex from love," and "whistles and wolf-calls, attendance at burlesque shows, [and] patronizing of call girls and prostitutes" are all "probably manifestations of a sexual urge totally or largely bereft of romantic feelings."

Argument 2 *Rape is not sexually motivated, because "most rapists have stable sexual partners" (Sanford and Fetter 1979, p. 8).*

This argument hinges on the assumption that a male's sexual desire is exhausted by a single partner. In addition to being contrary to our knowledge of the evolution of human sexuality, this assumption is obviously inconsistent with Symons's observation (1979, p. 280) that "most patrons of prostitutes, adult bookstores, and adult movie theaters are married men, but this is not considered evidence for lack of sexual motivation."

Argument 3 *Rape is not sexually motivated, because rapes are often "premeditated" (Brownmiller 1975; Griffin 1971).*

This argument hinges on the assumption that all acts that are truly sexually motivated are spontaneous. The assumption is obviously untrue: many highly planned affairs, rendezvous, and seductions are considered to be sexually motivated (Symons 1979, p. 279).

Argument 4 *The age distribution of rapists demonstrates that rape is a crime of violence and aggression rather than a crime of sex: ". . . the vio-*

lence prone years for males extend from their teenage years into their late forties, this is the age range into which most rapists fall. Unlike sexuality, aggression does diminish with age and, therefore, a male's likelihood of committing a rape diminishes with the onset of middle age." (Groth and Hobson 1983, p. 161)

Contrary to this assertion, the peak age distribution of rapists (teens through twenties; see Thornhill and Thornhill 1983) is perfectly consistent with the view that rapists are sexually motivated, since it closely parallels the age distribution of numerous other types of male sexual activity and of maximum male sexual motivation in general (Kinsey et al. 1948; Goethals 1971).[8]

Argument 5 *The fact that rape is common in war demonstrates that rape is motivated by hostility rather than sex (Brownmiller 1975, pp. 31–113; Card 1996).*

The high frequency of rape during war does not necessarily indicate that the rapists are not sexually motivated. The exceptionally high *vulnerability* of females during war may account for the greater frequency of rape by sexually motivated men. Theft is also frequent during war situations, owing to the fact that punishment is unlikely (Morris 1996), but this does not imply that the thieves are not motivated by desire for the stolen objects. Furthermore, the patterns of rape during war are consistent with the view that the rapist soldiers are sexually motivated and inconsistent with the view of rape as simply a tool of political domination. Throughout recorded history, the pattern in large-scale warfare has been to spare and rape the young non-pregnant women and to slaughter everyone else (Shields and Shields 1983; Hartung 1992).[9] Brownmiller (1975) sees rape in large-scale war as stemming in part from the frenzied state of affairs and the great excitement of men who have just forcefully dominated the enemy. That hypothesis predicts that soldier rapists would be indiscriminate about the age of the victims. But they are not; they prefer young women. Similarly, Brownmiller's view that rape in war—like rape in general—is a strategy of men to dominate women predicts that men would rape older women, who tend to have more resources and more social dominance.

***Argument* 6** *Rather than a sexually motivated act, rape is a form of "social control" because it is used as a form of punishment in some societies (Brownmiller 1975, p. 285).*

The flaw in this argument is that the use of rape as a punishment "does not prove that sexual feelings are not also involved, any more than the deprivation of property as punishment proves that the property is not valuable to the punisher" (Symons 1979, p. 280).

***Argument* 7** *"Men have been asked why they raped and many have said it was not out of sexual desire but for power and control over their victims." (Dean and de Bruyn-Kopps 1982, p. 233, citing evidence from Groth 1979)*[10]

Numerous studies have found that rapists often cite sexual desire as a cause of their actions. For example, Smithyman (1978, p. ix) reports that 84 percent of rapists surveyed cited sexual motivation "solely or in part" as a cause of their acts.[11] Indeed, even the quotations Groth (1979, pp. 38, 42) selected in an attempt to demonstrate the insignificance of sexual motivation includes such statements as "She stood there in her nightgown, and you could see right through it—you could see her nipples and breasts and, you know, they were just waiting for me, and it was just too much of a temptation to pass up" and "I just wanted to have sex with her and that was all."[12] Indeed, Groth (p. 28) points out that the most common type of rapist—what Groth calls the "power rapist"—"may report that his behavior was prompted by a desire for sexual gratification."

It is also important to note that reports of rapists' citing power and control rather than sexual desire as the cause of their actions come primarily from studies of *convicted* rapists. Were these men truthfully reporting their motives, or were they giving the explanations desired by the researchers? As Symons (1979, p. 283) observes, "it is difficult to avoid the conclusion that the men's conscious attempts to emphasize their correct attitudes and to minimize their sexual impulsiveness were to some extent calculated to foster the impression that they no longer constituted a threat."

Argument 8 *The physical harm done to victims demonstrates that rapists are not motivated by sexual desire (Harding 1985).*

To determine the significance of data on rapist violence and victim injury, one must distinguish between *instrumental* force (the force actually needed to complete the rape, and possibly to influence the victim not to resist, not to call for help, and/or not to report the rape) and *excessive* force (which might be a motivating end in itself). Only excessive force is a possible indication of violent motivation. Use of forceful tactics to reach a desired experience does not imply that the tactics are goals in themselves (unless, as was noted above, one is willing to argue that a man's giving money to a prostitute in exchange for sex is evidence that the man's behavior is motivated by a desire to give away money). Here again the crucial distinction between goals and tactics is blurred when rape is referred to as an act of violence.

Harding (1985, p. 51) claims that "in many cases of rape in humans, assault seems to be the important factor, not sex," on the ground that "in most cases the use of force goes beyond that necessary to compel compliance with the rapist's demands." However, it is evident from the actual data—including the data that Harding cites on the very same page—that, although force is often used instrumentally to accomplish a rape, excessive force resulting in substantial physical injuries occurs only in a minority of rapes. In their study of 1401 rape victims, McCahill et al. (1979) found that most of the victims reported the use of instrumental force (84 percent reported being threatened with bodily harm, 64 percent being pushed or held), but acts that might indicate excessive force were reported in only a minority of the rapes (slapping in 17 percent, beating in 22 percent, choking in 20 percent). Similarly, a survey of volunteers at rape crisis centers found that only 15 percent of victims they encountered reported having been beaten in excess of what was needed to accomplish the rape (Palmer 1988b, p. 219).[13] Geis (1977) found that 78 percent of the rapists in his study had wanted the victim to cooperate.[14] Katz and Mazur (1979, p. 171) found that "although most rape victims encountered some form of physical force, few experienced severe lasting injuries"—a pattern also reported by Bowyer and Dalton (1997). Even a study that focused on "overly violent rapists" (Queen's Bench Foundation 1978, p. 778) found that only 23 percent of these rapists inflicted "very severe injury." In comparison with Harding's assertion, the evidence appears to be more consistent with Hagen's (1979, p. 87) conclusion:

> . . . in the great majority of rape cases, physical injury, other than that which might be related to penetration is not done to the victim (for example Brownmiller 1975, p. 216; Burgess and Holmstrom 1974). And generally, there is no injury at all. If violence is what the rapist is after, he's not very good at it. Certainly he has the victim in a position from which he could do all kinds of physical damage.

Even when excessive violence does occur, sexual motivation still appears to be a necessary part of the explanation for why a rape rather than a nonsexual assault occurred. As Rada (1978a, p. 22) states, "if aggression were the sole motive it might be more simply satisfied by a physical beating."

Although murder of a rape victim certainly may indicate hostile motivation, at least some such murders may be due to the simple fact that killing the victim greatly increases the rapist's chances of escaping punishment by removing the only witness to the rape (Alexander and Noonan 1979; Groth 1979; Hagen 1979). Rape-murders, however, are a very small percentage of all murders. In the United States, over the period 1976–1994, in no year was the percentage of murders that included rape or other sexual assault higher than 2 (Greenfield 1997)—and an unknown portion of that small percentage involved *male* murder victims.

An evolutionary approach can also explain patterns of excessive force in the minority of cases where it does occur. Young women, highly overrepresented as rape victims, are also at the greatest risk of being killed by their assailants, according to data from the United Kingdom, Canada, and Chicago (Wilson et al. 1997). Young women appear to resist rape more than females in other age groups. The strong sexual motivation of the rapist to rape a young victim, in combination with her greater resistance, may account for young women's overrepresentation in homicides with sexual assault. And female victims of theft-murder are, on average, much older than female victims of rape-murder (Wilson et al. 1997).

Another circumstance that is probably related to the use of excessive violence by some rapists is dissolution of mateship. Men's sexual jealousy and other proprietary actions toward deserting mates often includes battering as a mechanism of sexual control. Another abusive behavior may be inseminating the deserting mate against her will, which has the (perhaps evolved) effect of increasing the rapist's paternity reliability if there is sperm competition. There are data indicating that men who rape their estranged mates are more likely to physically injure the victim than rapists who have other relationships to the victim (Felson and Krohn 1990).

While contrary to the social science explanation of rape, evidence that rapists do not routinely use excess violence in order to mate with unwilling women is predicted by evolutionary theory. Rape occurs against the will of the victim and thus is often accompanied by tactical violence. However, violence that injures the victim would reduce her ability to produce and care for any offspring that resulted from the rape. This cost, which applies to much of human evolutionary history, is expected to have given rise to selection for rapists who minimize injury to their victims.[15]

As was detailed in chapter 3, the evolutionary view of rape as rape-specific adaptation suggests that men may be sexually aroused by physical control of the victim because such control would have facilitated rape in human evolutionary history while also reducing the cost of rape to the rapist. This does not imply that rape motivation of men requires physical control of the victim; it implies only that such control, when perceived by men, may increase rape motivation because it increases their sexual arousal.

Argument 9 *IT IS NOT A CRIME OF LUST BUT OF VIOLENCE AND POWER . . . RAPE VICTIMS ARE NOT ONLY THE "LOVELY YOUNG BLONDES" OF NEWSPAPER HEADLINES—RAPISTS STRIKE CHILDREN, THE AGED, THE HOMELY—ALL WOMEN. (Brownmiller 1976, back cover)*

It is fitting that this argument should appear in bold type on the cover of Brownmiller's milestone book *Against Our Will*, since the assertion that rapists do not prefer sexually attractive victims is probably the most powerful and the most widely cited argument used to support the claim that rapists are not sexually motivated (Palmer 1988a). That argument is fatally flawed, however. The statement that "any female may become a victim of rape" (Brownmiller 1975, p. 348) does not imply that the "rapist chooses his victim with a striking disregard for conventional 'sex appeal'" (ibid., p. 338). Contrary to Brownmiller, although any female *might* become a victim of rape, some women are far more likely to become victims of rape than others. Indeed, one of the most consistent finding of studies on rape, and one not likely to be due entirely to reporting bias, is that women in their teens and their early twenties are highly overrepresented among rape victims around the world (Svalastoga 1962; Amir 1971; MacDonald 1971; Miyazawa 1976; Hindelang 1977; Hindelang and Davis 1977; Russell 1984; Kramer 1987; Whitaker 1987; Pawson and Banks 1993). Far from demonstrating the absence of sexual motivation in rapists, the correlation between the age distribution of rape victims and the age of peak female sexual attractiveness is powerful evidence of such motivation. Other such evidence is provided by the finding that during sexual assaults men are more likely to engage in penile-vaginal intercourse (as opposed to exclusively non-copulatory sexual behaviors), and in multiple episodes of such intercourse, when the victim is a young woman than when she is of non-reproductive age. (See chapter 4.)

Many of the researchers who have denied the importance of sexual motivation in rape have asserted that the vulnerability of victims is the primary factor explaining the age pattern of victimization. For example, Groth (1979, p. 173) states that "vulnerability and accessibility play a more significant role in determining victim selection than does physical attractiveness or alleged provocativeness" and that "rape is far more an issue of hostility than of sexual desire." This argument is truly astonishing in view of the fact that the age distribution of rape victims is essentially the opposite of what would be predicted by this explanation. Numerous researchers, including Groth, have pointed out that females in the age categories *least* likely to be raped are the *most* vulnerable. As Rodabaugh and Austin (1981, p. 44) note, "both the very young and the very old [are] at high risk because of their inability to resist." Indeed, although the elderly are "particularly vulnerable" to rape (Groth 1979, p. 173), various studies, including Groth's own, have consistently found that less than 5 percent of rape victims are over the age of 50. Although the greater vulnerability of children and the elderly probably accounts for why they are raped more often than would be expected on the basis of their attractiveness alone, they are raped at a rate far below what would be expected if vulnerability were the only factor or even the primary factor in victim selection.

Skeptical review of all these arguments supports Paglia's (1994, p. 41) assertion that "the rape discourse derailed itself early on by its nonsensical formulation, 'Rape is a crime of violence but not of sex,' a mantra that blanketed the American media."

Cross-Cultural Evidence

The other basic premise of the social science theory of rape is that rape occurs only when it is taught, or encouraged in some other way, by a culture. Although it would be logically possible for this explanation to be correct and for rape to occur in all cultures, that would require the assumption that, just by coincidence, all cultures encourage males to rape. Owing to the extreme improbability of such a coincidence, most social science discussions of rape in a cross-cultural context emphasize the variability of rape's occurrence, and even its absence in some cultures—an emphasis that is consistent with the social science view that socialization is arbitrary. In addition, the view of rape as the product of only certain cultures, including the researcher's own culture, seems to offer useful direction for social reform in the opinion of those with the view. If rape is the product of only certain aspects of our culture, then we ought to change those aspects.

Griffin (1971) asserts that rape is absent from some cultures and cites as evidence Margaret Mead's (1935) statement about the Arapesh. From this alleged lack of universal occurrence, Griffin concludes that "far from the social control of rape being learned, comparisons with other cultures lead one to suspect that, in our society, it is rape itself that is learned" (p. 28). This conclusion about the cause of rape then leads to the solution of changing those cultures, such as Griffin's own, in which rape occurs. The nature of those changes is predictable in view of the connection between "science" and ideology in the social sciences: "Rape is not an isolated act that can be rooted out from patriarchy without ending patriarchy itself." (ibid., p. 36)

Supposed proof of both the lack of universality of rape and Griffin's hypothesized connection between patriarchy and rape was put forth by

Sanday (1981), who claimed to have found 45 out of the 95 societies in her sample to be "rape free." Further, Sanday claimed to have found that rape occurs primarily in patriarchal societies that are out of touch with nature. The similarity between these "findings" and feminist ideological values is probably not coincidental. In any case, Sanday's characterization of 45 of the 95 of the societies in her sample as "rape free" is clearly inaccurate (as she admits on page 9 of her article), since she includes in the "rape free" category societies where rapes are supposedly "rare." In actuality, Sanday's own descriptions indicate the presence of rape in all but five of the cultures in her sample.

Furthermore, the ethnographic evidence (Palmer 1989a) does not justify the claim of the absence of rape in the five "rape-free" cultures identified by Sanday, nor does it justify similar claims about other cultures (Broude and Greene 1976; Minturn et al. 1969).

Some assertions of the absence of rape in certain cultures may be due to incomplete consideration of ethnographic data; in other cases, however, it is difficult to avoid the possibility that other motives were involved. For example, Sanday (1981, p. 17) states that "Turnbull's [1965] description of the Mbuti Pygmies, of the Ituri forest in Africa, provides a prototypical profile of a 'rape free' society." Sanday bases this claim on the relative lack of patriarchy among the Mbuti and on her assertion that "Turnbull . . . , an anthropologist who lived for sometime among the Pygmies and became closely identified with them, reports that he knew of no cases of rape" (p. 16). However, Turnbull's actual statement (1965, p. 121) reads as follows: "I know of no cases of rape, though boys often talk about their intentions of forcing reluctant maidens to their will." On page 137 of the same book, Turnbull reports that during the *elima* (a female initiation ceremony), although the rules state that a male "has to have [the girl's] permission before intercourse can take place," in reality "the men say that once they lie down with a girl, . . . if they want her they take her by surprise when petting her, and force her to their will."[16]

Mead's famous description of the Arapesh fares no better than Turnbull's description of a "rape-free" society. Yet not only is Mead's assertion one of the original pillars of the feminist–social science explanation of rape; it is, according to Geis (1977, p. 30), "undoubtedly the most widely quoted ethnographic remark on rape."

Mead's treatment of the Arapesh begins as follows (1935, p. 104): "Of rape the Arapesh know nothing beyond the fact that it is the unpleasant custom of the Nugum people to the southeast of them." Mead's subsequent statement appears to support the social science explanation perfectly: "Nor do the Arapesh have any conception of male nature that might make rape understandable to them." To support that statement, Mead then offers the following: "If a man carries off a woman whom he has not won through seduction, he will not take her at once, in the heat of his excitement over having captured her. Rather he will delay soberly until he sees which way negotiations turn, whether there is a battle over her, what pressure is brought upon him to return her. If she is not to belong to him permanently, it is much safer never to possess her at all." Though this is intended to support her claim that the Arapesh males find rape incomprehensible, the behavior Mead describes is rape: Arapesh males forcibly abduct non-consenting women for sexual intercourse, and they complete the rape whenever the consequences of the act are not expected to be too severe.

The ethnographic evidence indicates that some frequency of rape is typical of *Homo sapiens* and that there is no evidence of a truly rape-free society (Palmer 1989a; Rozée 1993; Wrangham and Peterson 1996; Jones 1999). This does not mean, of course, that rape is a genetically determined act unaffected by learning and culture. It means only that human males in all societies so far examined in the ethnographic record possess genes that can lead, by way of ontogeny, to raping behavior when the necessary environmental factors are present, and that the necessary environmental factors are sometimes present in all societies studied to date.

Further, the actual role of the specific environmental influence of other individuals (commonly referred to as "culture") on the development of male sexuality appears to be far different from the role assigned to it by the social science model. The social science model holds that experiencing other individuals' explicit or implicit encouragement of raping behavior is a necessary precursor to rape. That is, "it is rape itself that is learned" (Griffin 1971, p. 28). Though there is little doubt that such encouragement of rape may increase the chances of its occurrence, cross-cultural data clearly demonstrates that such encouragement is far from necessary.

Close examination of ethnographic data on the cultures in which it has been asserted that rape is accepted and never punished (Broude and Greene 1976; Minturn et al. 1969; Sanday 1981) reveals all such claims to be unfounded (Palmer 1989a). Indeed, although rape may be accepted or even encouraged in certain restricted situations in some cultures (outgroup rape is accepted or encouraged in wartime in some cases; see, e.g., Shields and Shields 1983), in the cultures examined some forms of rape appear to occur despite being punished. That is, rape occurs under a much broader set of environmental influences than those proposed by social scientists, and, contrary to the social science model, comparisons with other cultures indicate that it is the "social control of rape," not rape itself, that is most often encouraged by the influence of others (that is, socially learned).

Cross-Species Evidence

The social science explanation of rape asserts that socialization alone causes the sex differences that produce rape: males, relative to females, are more aggressive, more sexually assertive, more eager to copulate, and less discriminating of mates. However, these same sex differences occur in all animals whose evolutionary history involves polygyny and sexual disparity in the minimum cost of offspring production. Moreover, the vast majority of these species (including all polygynous invertebrates, comprising millions of species) show no sexual training of juveniles by other group members, and none show the extensive sexual socialization seen in humans. Thus, the sex differences in sexual socialization seen in humans cannot be viewed parsimoniously as the only significant cause of the basic sex differences in human sexuality, since the same pattern of sex differences is evidently universal in polygynous species. Across species, the "common denominator" in this pattern is an evolutionary history that involves greater competition among males than among females for sexual access to multiple mates, not human-like sexual socialization. Sexual socialization is essentially irrelevant to rape in the non-human species.

Indeed, evolutionary biology predicts rape in species with the abovementioned sex differences whenever the benefits of the act outweigh the costs to males. (As always in evolutionary biology, 'benefits' and 'costs' refer to reproductive consequences in the evolutionary historical environments of species.) On the other hand, widespread occurrence of rape in non-human species is completely incompatible with the social science explanation. Perhaps realizing this, Brownmiller (1975, p. 12) stated the following: "No zoologist, as far as I know, has ever observed that animals rape in their natural habitat, the wild." Actually, there were already a considerable number of published evolutionary analyses of sexual coercion in non-humans (Severinghaus 1955; Barlow 1967; Manning 1967; Van Den Assem 1967; Fishelson 1970; Lorenz 1970, 1971; Keeneyside 1972; Linley 1972; Pinto 1972; MacKinnon 1974; Parker 1974).

In the ten years that followed Brownmiller's claim, studies of rape in non-human species grew too numerous to be ignored. Evolutionary explanations of rape were put forth in regard to insects (Las 1972; Oh 1979; Parker 1979; Cade 1980; Pinkser and Doschek 1980; Smith and Prokopy 1980; Thornhill 1980, 1981, 1984; Thornhill and Alcock 1983; Crespi 1986; Tsubaki and Ono 1986), birds (Afton 1985; Hoogland and Sherman 1976; Barash 1977; Bailey et al. 1978; Beecher and Beecher 1979; Birkhead 1979; Gladstone 1979; Mineau and Cooke 1979; Bingman 1980; Burns et al. 1980; McKinney et al. 1980; Seymour and Titman 1980; McKinney and Stolen 1982; Cheng et al. 1983a,b; Titman 1983; Birkhead et al. 1985; Bossema and Roemers 1985; Van Rhijn and Groethuis 1985; Emlen and Wrege 1986), fishes (Constantz 1975; Kodric-Brown 1977; Farr 1980; Farr et al. 1986), reptiles and amphibians (Wells 1977; Howard 1978; Cooper 1985), marine mammals (Cox and Le Boeuf 1977), and non-human primates (Rijksen 1978; Galdikas 1979, 1985a,b; MacKinnon 1979; Nadler and Miller 1982; Jones 1985; Mitani 1985; Goodall 1986).

Research on rape in non-human species continues to be regularly reported in biological journals and at meetings of ethologists (Thornhill 1987; Arnqvist 1989, 1992; Mesnick and Le Boeuf 1991; Thornhill and Sauer 1991; Thornhill 1992a,b; Hemni et al. 1993; Smuts and Smuts 1993; Sorenson 1994; Arnqvist and Rowe 1995; Clutton-Brock and Parker 1995; Sakaluk et al. 1995; Allen and Simmons 1996; Andersen 1997; Soltis et al. 1997; many others). There is no longer any question that physical force, harassment, and intimidation are used widely by males across animal species, including in the great apes, to obtain mates (Smuts and Smuts 1993; Clutton-Brock and Parker 1995; Wrangham and Peterson 1996; Nadler 1999).

Rape is especially common in the orangutan. The accumulation of data on orangutans' sexual behavior in nature by a group of dedicated researchers has led to the tentative conclusion that orangutan males may exist as two distinct morphs (Wrangham and Peterson 1996). Males of the large morph, weighing about 90 kilograms, move slowly through the canopy of rainforest trees and are attractive to females, who mate willingly with them. Males of the small morph are about the same size as females (40 kg), travel as fast in the trees as the females, tend to be avoided by females, and run females down for rape. There are three indications that the two types of adult males may be two tactics of one conditional strategy that is possessed by all male orangutans: First, the small males occasionally "undergo a sudden growth spurt and turn into big males" (ibid., p. 135). Second, there is some evidence from captivity that small males remain small when there is a large male in the vicinity. Third, unlike the big males, the small males avoid male-male fights.

The social system of the orangutan is quite different from that of the other apes in that all individual orangutans live alone. Thus, all female orangutans lack pair-bond mates or kin who might thwart rape attempts, which according to Wrangham and Peterson (1996, p. 142) appear to account for "one-third to one-half or more of all copulations." Wrangham and Peterson speculate that variation in the element of protection across ape social systems may account for different frequencies of sexual coercion that have been found in the apes.[17] Social-system variation also seems to be important in understanding the evolution of rape and other forms of sexual coercion in other primates and in certain other animal groups (Clutton-Brock and Parker 1995).

Rape has been observed in more than 39 species of pair-bonding birds (McKinney et al. 1983; Sorenson 1994). In the common mallard duck, males guard their mates from rape while the mate is fertile and then neglect them to attempt rapes of other fertile females while their infertile mates incubate eggs (Barash 1977; Evarts 1990, cited and discussed in Sorenson 1994).

Across many species, it is not maleness itself that results in sexual coercion by males and thus in the insemination of more females. Rather, rape and other forms of sexual coercion show an association with males because, typically, males exhibit less parental effort than females and thus have a history of stronger selection for high mate number. As Smuts and Smuts (1993, p. 44) point out, "female sexual coercion is expected to occur . . . in sex-role reversed species, in which females compete intensively for mating opportunities with males." In these species, males require copulation before they will invest their disproportionately greater amount of parental effort. Effective selection on females for coercing mating may be a possibility if this gives access to more male investment. In seahorses, the female has the "penis" and thus rape by females is a possibility.

The widespread occurrence of rape across animal species is both consistent with evolutionary predictions and devastating to the social science explanation. This is apparently why social scientists, faced with overwhelming evidence from non-human species, abandoned Brownmiller's assertion that non-human species did not exhibit rape behavior and adopted the term "forced copulation" for rape in non-human species. Because cross-species comparisons are a critical source of tests of ultimate hypotheses (Williams 1992; Alcock 1997), this semantic evasion can only reduce people's understanding of the causes of rape.

Researchers have just begun to examine the cross-species presence and absence of rape and other sexual coercion in the light of ecological variables that are hypothesized to affect female vulnerability and selection pressures.

Smuts (1992) has studied several variables that may be relevant to the conditional use of sexual coercion by men across human cultures: protection of women by kin, guarding of mates by males, male-male political alliances, and males' control of resources. Although there is as yet only a limited understanding of how these variables correlate with degree and type of sexual coercion across human groups, this approach emphasizes the conditional nature of rape.

Metaphysical Assumptions

Aside from its empirical and logical flaws, the social science theory of rape hinges on two metaphysical assertions that remove it from the realm of science: the causality of the cultural spirit and the dichotomy of mind and body.

First, the theory attributes cause to a non-corporeal reified entity referred to as "a culture" or "a society." In 1972, around the time that the idea of group selection as an effective force responsible for adaptation was being overthrown in biology, the anthropologist George Peter Murdock stated the following:

> It now seems to me distressingly obvious that culture, social system, and all comparable supra-individual concepts, such as collective representations, group mind, and social organism, are illusory conceptual abstractions inferred from observations of the very real phenomena of individuals interacting with one another and with their natural environments. . . . They are, in short, mythology, not science, and are to be rejected in their entirety—not revised or modified.[18] (Murdock 1972, p. 19)

Despite Murdock's strong advice, many social science explanations, including explanations of rape, continue to attribute proximate causation to abstract and metaphysical group entities.

The second premise of the social science theory of rape—that rape is not sexually motivated—also contains a metaphysical assumption. The claim that sexual arousal, interest, and/or motivation is absent during sexual acts implies an extreme form of the classic dualistic assumption that human brains (or minds) are separate entities from bodies—a notion long ago tossed on the intellectual trash heap. For example, Sanday (1990, p. 10) states that during a rape "the sexual act is not concerned with sexual gratification but with the deployment of the penis as a concrete symbol of masculine social power," and Beneke (1982, p. 16) asserts that, for males, not only does rape have nothing to do with sex; "sex itself often has little to do with sex." This implies that the bodies of human males may go through all of the physiological processes of sex (arousal, erection, even ejaculation) without their brains' going through corresponding sexual physiological processes (such as dopamine reward). Since certain physiological processes in the brain can be shown to accompany certain physiological processes in the rest of the body, the alleged lack of sexual motivation in the brain during sexual acts must refer to the state of an unidentifiable human mind distinct from the brain.

The Empirical Product of the Social Science Explanation

The evolutionarily informed psychologists Del Thiessen and Robert Young (1994) investigated 1610 studies of human sexual coercion published between 1982 and 1992. Their sample included studies done by psychologists, by educational psychologists, by anthropologists, and by sociologists. They found that fewer than 10 percent of the studies were directed at understanding the causes of sexual coercion, that hypotheses were tested in only 9 percent of the studies, that only about 9 percent of the studies showed any sign of quantification, and that in exactly 1.5 percent of the studies was a statistical test applied. Thiessen and Young also found no significant changes in the foci of studies or in the conclusions between the period 1982–1987 and the period 1987–1992, whereas truly scientific endeavors are characterized by increasing refinement of hypotheses based on the rejection of previous hypotheses that failed quantitative tests.

Not only is the bulk of the social science literature of rape clearly indifferent to scientific standards; many of the studies exhibit overt hostility toward scientific approaches, and specifically toward biological approaches. The message of these studies is clearly political rather than scientific. Many of the social science studies that Thiessen and Young investigated focused on "consciousness raising" and blamed social policies and male oppression for rape victims' problems. As Thiessen and Young emphasize, responsibility for the sad state of affairs in rape research must be distributed among investigators, journal editors, funding agencies, academic departments, and university administrators.

Ideology and the Social Science Explanation

Despite its incompatibility with the basic principles of biology and its failure to generate cumulative scientific knowledge, the current social science theory of rape has been popular for more than 25 years. Several reasons have been proposed for why such an unsupported position could have achieved and maintained such popularity.

Perhaps the naturalistic fallacy leads social scientists to fear that if rape is motivated by something as natural as sexual desire then it must be good, or at least excusable. Symons (1979) suggests that the "not sex" argument is attributable to the view that sex is good and therefore cannot be involved in something bad. Another possibility, suggested by Thornhill and Thornhill (1983), is that viewing rape as distinct from sex is attributable to the importance of female choice throughout human evolutionary history. If female sexual arousal has been designed to occur when a female finds a male who has the traits and the behavior of a good mate, then rape is exceedingly unlikely to be sexually arousing to females. Thus, for females, the experience of rape is very different from that of sex with a desired mate.

However, the main reason for the denial that rapists are sexually motivated almost certainly stems from political ideology. Debates about what causes rape have been evaluated not on the basis of logic and evidence but on the basis of how the different positions might influence people to behave. Consistent with this analysis is the fact that ideology has always played a role in support for the "not sex" explanation of rape. "The 'rape as violence' position," Estrich (1987, p. 82) states, "has always seemed to me the better approach both theoretically and strategically." Muehlenhard et al. (1996, p. 123) observe that "feminist theorists . . . argued for the utility of conceptualizing rape as violence." Scully and Marolla (1995, p. 66) state that "in an effort to change public attitudes that are damaging to the victims of rape . . . many writers . . . discount the part that sex plays in the crime." MacKinnon (1990, p. 5) characterizes the feminism that put forth the "not sex" argument as "a movement that took women's side in everything."

The main reason the "not sex" explanation of rape was seen as good for women is that the fallacy of genetic determinism causes sexual desires to be mistakenly equated with uncontrollable lust. As Symons (1979, p. 279) notes, "many writers seem to fear that to admit sex as a motive for rape is to risk condoning rape: lust is presumed to be less easily controlled through an act of will than are other possible motives for rape, hence, in this view, if lust motivates rape, the rapist cannot be held fully accountable for his actions."

Ideological considerations about how the world should be, much more than evidence about how the world is, are also behind the minor deviations from the "not sex" explanation of rape that have arisen among feminists. Some authors have argued that rapists are motivated by a desire for both sex and violence, not necessarily because the empirical evidence demonstrates this statement to be true, but because they expect that this explanation will have better consequences than the "not sex" explanation. Consider the following: "It has sometimes been assumed that conceptualizing rape as sex has solely negative implications for women, whereas conceptualizing rape as violence has solely positive implications. We believe, however, that the situation is considerably more complex." (Muehlenhard et al. 1996, p. 130)

Concerns about the practical consequences of explanations also appear to be behind the rejection of the "rape as violence" position in favor of the "rape as sexualized violence" position put forth by radical feminists. "Radical feminists," notes Torrey (1995, p. 44), "assert that it is a mistake to characterize rape as violence rather than sexualized violence. As a practical matter, prosecuting rapes that may lack physical violence, such as date and marital rape, is more difficult under a characterization of rape as simply violence."

The preeminence of ideology is also reflected in recent discussions of the *definition* of rape. Following Estrich (1987), we have used the word 'rape' to refer to human copulation resisted by the victim to the best of her ability unless such resistance would probably result in death or serious injury to her or in death or injury to others she commonly protects.[19] We use this definition because it is consistent with how the word is used by most people. In particular, it distinguishes copulations referred to as rapes from other copulations involving coercion. We use 'coercion' to mean any form of force or influence that involves application or threat of a negative consequence. The consequences associated with sexual coercion may range from implicit threats of withdrawal of cooperation or emotional involvement to death. Although all these interactions may be of interest, they are not all literally referred to by the word 'rape.'

When ideology and the naturalistic fallacy rule, the definition of a word becomes merely a tool used to persuade people to adopt a certain position about how the world ought to be. Such uses of definitions are sometimes justified on the grounds that definitions are arbitrary and the notion of a true definition is "a meaningless concept" (Muehlenhard et al. 1996, p. 124). The assertion that there are no accurate definitions, however, implies that there are also no inaccurate definitions—an assertion that is obviously false. More fundamentally, there would be no reason to write the statement "there is no such thing as an accurate definition" if the writer did not assume that the reader would share the writer's understanding of the definitions of those words. Hence, to write that "there is no such thing as an accurate definition" is to engage in a self-contradictory act.

As Muehlenhard et al. (1996, p. 124) point out, certain interest groups sometimes "develop their own definitions" as an "act of resistance." When this happens, definitions are evaluated not for their consistency with how the word in question is used generally but for their political "utility" in influencing people to adopt some desired pattern of behavior. This is much in evidence in many feminist discussions of rape, which often are based on "definitions of rape that are much broader than traditional definitions" (Muehlenhard et al. 1996, p. 125). For example, several authors have asserted that all or nearly all heterosexual acts are rapes (MacKinnon 1987; Southern Women's Writing Collective 1990). "At the liberal extreme," writes Bourque (1989, p. 6), "any sexual behavior, even off-color jokes at work, unsolicited use of diminutives, or a hand on the arm, constitutes rape if a woman indicates by word or deed that such actions impinge on her personal space." The political motives behind such assertions are clear in the following statement by MacKinnon (1987, p. 82): "Politically, I call it rape whenever a woman has sex and feels violated." Though such metaphorical uses of the word 'rape' (and even assertions that clear cutting of timber rapes a forest) are appropriate as metaphors to the extent that the activity in question is *like* rape, these activities clearly are not rape. That is, they are not literally rape, and in reality most people are able to distinguish these activities from the activity they literally call rape.

Blackman (1985, p. 118) criticizes the evolutionary approach for "de-politicizing" the word 'rape' on the ground that "to use the word . . . in a de-politicized context functions to undermine ten years of feminist consciousness-raising." However, clinging to unsupported, contradicted explanations of the causes of rape actually hinders attempts to prevent rape. Indeed, successfully reducing the incidence of rape may require that rape be de-politicized from its status as the "master symbol of women's oppression" (Schwendinger and Schwendinger 1985, p. 93) to the status of a behavior to be prevented through the identification of its causes. Such a de-politicization is particularly crucial in regard to the issue of sexual motivation, because "to the extent that men's sexual arousal to rape cues precipitates rape, assessment and treatment of such sexual arousal may help prevent rape" (Muehlenhard et al. 1996, p. 130).

The full extent of the role of ideology in determining the acceptance of the social science theory of rape can be seen by considering reactions to a hypothetical explanation based on the very same assumptions about human nature that form the social science theory, but with very different ideological connotations. The implausibility of an infinitely flexible human nature becomes obvious when explanations contrary to one's ideology are proposed. The social science explanation of rape clearly implies that women find rape a negative experience only *when they are influenced by their culture to feel this way*. If this were true, then stopping rape would not be necessary in order to solve rape as a social *problem*. Instead, according to the assumptions of the social science explanation of rape, the *problem* of rape could be solved simply by teaching women that rape is a wonderful experience. If this course of action sounds absurd (and it sounds very absurd to us—see chapter 4), it is because the assumption on which it is based is so implausible. Human females obviously do not have a nature so flexible that they could come to desire the experience of being raped simply by being educated to do so. Yet this is exactly the flexible human nature implied by the social science theory of rape. The ability of ideology to blind people to the utter implausibility of their positions is perhaps the greatest threat to accumulating the knowledge necessary to solve social problems.

7

Law and Punishment

According to the social science model, rape is culturally determined, not genetically determined. Cultural determinism is consistent with free will and with the ability of humans to change their behavior easily by adopting new social constructs. This model is in conflict with everything that is known about the interaction of genetic and environmental factors in the development of all behavioral abilities and about the effects of selection on the shaping of all the adaptations involved in behavioral development.

The evolutionary approach holds that no behavior is inevitable. Only by understanding this point can we hope to understand how human-mediated alterations in the developmental environment can produce desirable behavioral changes. Knowledge about evolution is required if we are ever to escape from what Symons (1979, p. 313) called "the nightmare of the past." With respect to rape, the power of the evolutionary approach lies in its ability to identify environmental changes that may remove cues that activate the evolved mechanisms that underlie rape behavior.

Could rape be eliminated from the human population by eugenic means—that is, by selecting against impulsiveness and psychopathy in a manner analogous to the artificial selection of desirable traits in livestock and poultry?[1] Behaviors that might be selected *for* are indiscriminate mating by females (so that rape wouldn't matter to them) and, in males, sexual interest only in committed relationships (so that male sexual arousal wouldn't occur in the absence of commitment). Of course, a host of practical, ethical, and historical issues would rule out such a program, and in any case such a program would take far too long. For selection to create complex adaptation takes hundreds or even thousands of generations. And working against artificial selection's effectiveness is counter-selection—for example, the same genes that are associated with greater sexual impulsiveness may (at least in some circumstances) confer a hidden advantage, such as resistance to infectious disease. Thus, for both moral and practical reasons, the idea of using artificial selection to solve the problem of rape can be ignored.

A far more practical and far more moral approach—an approach based on knowledge of the ontogeny of behavior—would address environmental factors. For example, encouraging the rearing of boys in environments rich with enduring personal relationships (and, in particular, with the father present) might well reduce the development of the proclivity to rape. Such an approach would require much more research dedicated to discovering the key developmental cues. Evolutionary theory would be crucial, since it predicts that the developmental events of interest will occur in response to specific cues that, in our history as a species, were most reliably correlated with reduced consensual sex with females.

Related to the developmental approach is the use of evolutionary theory to identify the proximate cues that activate the psychological adaptations responsible for rape after their developmental construction. That most men don't rape is indicative of the potential fruitfulness of this approach. It is now a matter of determining what the relevant cues are. Evolutionary theory points to the need to discover the factors that affected the benefits and costs of rape to adult males in human evolutionary history.

Contrary to the common view that an evolutionary explanation for human behavior removes individuals' responsibility for their actions, individuals who really understood the evolutionary bases of their actions might be better able to avoid behaving in an "adaptive" fashion that is damaging to others. As Alexander (1979) has emphasized, knowledge of the self as having evolved by Darwinian selection provides an individual with tremendous potential for free will. Moreover, refusal to refrain from damaging behavior in the face of scientific understanding could be seen as a ground for holding irresponsible individuals *more* culpable, not less so. That rape is entirely based on biology does not imply that men cannot consciously choose not to rape.

Rape Laws

Foremost among social scientists' concerns about the laws regarding rape is the fact that rape has traditionally been defined and punished not from the victim's perspective but from a male perspective, and particularly from the perspective of the victim's mate. For example, White and Sorenson (1992, p. 190) state that traditionally "rape is defined from a heterosexual male perspective," and Berger et al. (1988, p. 330) write that rape laws "regulated women's sexuality and protected male rights to possess women as sexual objects." According to the legal scholars Cassia Spohn and Julie Horney (1992, pp. 21–22), "historically, rape was defined as 'carnal knowledge of a woman, not one's wife, by force and against her will,' and "carnal knowledge included only penile-vaginal penetration." Donat and D'Emilio (1992, p. 10) state that "a woman who was sexually attacked needed to comply with male standards for her behavior by proving her nonconsent," and that "proof of nonconsent was necessary to verify that the woman had not voluntarily engaged in sexual acts outside of marriage." "If a woman could not prove nonconsent," Donat and D'Emilio continue, "she might be punished for the assault." The perspective of the victim's mate is evident in the Talmud (a collection of laws derived from the Torah by rabbis called Sages) and in the Talmud's derivative, the Codes of Maimonides.[2]

Social scientists and legal scholars have also complained about the "pervasive skepticism of the claims of rape victims" (Spohn and Horney 1992, p. 18). For example, "to demonstrate her nonconsent, the victim was required under many statutes to 'resist to the utmost' or, at the very least, to exhibit 'such earnest resistance as might reasonably be expected under the circumstances'" (ibid., p. 23). That a victim's past sexual behavior is sometimes considered in regard to whether or not she consented and that "victims who know their attackers or who somehow 'precipitated' the attack by their dress, behavior, or reputation must prove that they are worthy of protection under the law" (ibid., p. 20) have also been criticized.[3]

In the 1970s and the 1980s a movement arose to change the aforementioned aspects of the rape laws. In addition to "redefining rape and replacing the single crime of rape with a series of graded offenses" (Spohn and Horney 1992, p. 21), reformers "criticized rules of evidence that required the victim to physically resist her attacker, that required corroboration of the victim's testimony, and that allowed evidence of the victim's past sexual conduct to be admitted at trial"; they also "criticized common-law definitions of rape that excluded males and spouses as victims and that excluded acts other that sexual intercourse" (ibid., p. 18).

The first goal of the movement to reform rape laws was to get lawyers and judges to abandon the traditional category of rape. However, a team of legal scholars reported in 1988 that prosecutors continued to "distinguish cases of 'real' or 'classic' rape from other sex offenses" (Berger et al. 1988, p. 334)—a practice that the scholars called "not consistent with reformers' goals of defining a continuum of offenses." Similarly, "the rape shield laws did not produce the changes envisioned by reformers" (Spohn and Horney 1992, p. 164), and it was reported that the assumption that past sexual behavior is relevant to the issue of consent had "not been substantially altered by a procedural change in the law" (ibid., p. 31). Furthermore, "reforms eliminating corroboration and resistance requirements . . . had little impact because corroboration and resistance evidence are still considered essential for obtaining a conviction"; for example, "inserting a statement that 'the victim need not resist the accused' into the statute does not preclude the prosecutor from taking the victim's lack of resistance into account when deciding whether to file charges or not" (ibid., pp. 162–163).

The reason the movement to reform rape laws has met with only limited success is that the reformers are trying to change attitudes toward rape in the absence of an understanding of the evolved psychological mechanisms that produce those attitudes. The evolutionary approach illuminates why there are rape laws in the first place and why certain characteristics of the existing laws seem so puzzling to those who fail to take the evolutionary perspective.

Humans distinguish rape from other copulation, and even from other forms of copulation that involve coercion (Palmer 1989a), because this particular form of copulation has had specific deleterious consequences to the reproductive success of individuals throughout human evolutionary history. Not only has it reduced the female victim's reproductive success; it has also reduced the fitness of the victim's kin, and especially that of her mate. Hence, rape has very likely been an obstacle to reproductive success that has led to adaptations in both females and males. The adaptations in the female psyche designed to avoid rape were discussed in chapter 4. Selection would have also favored males who reacted to this category of copulation in certain ways. We suggest that such specific "rape reaction" psychological mechanisms exist in human males, and that they explain many facets of rape laws. Indeed, we hypothesize that these male psychological adaptations are the main obstacles to attempts to reform rape laws.

The rape of a man's wife is a threat to the man's reproductive success because it threatens his paternity certainty. Hence, selection favored males who responded to the rape of their wives in certain ways. A male who discovered that his mate had copulated with another male was faced with deciding whether to abandon her or to continue to invest in her and her future offspring. For the male, the first option has the cost of abandoning existing offspring he may have sired with the mate and the cost of then having to find another mate; it has the benefit of not wasting investment on an offspring sired by a rapist. Hence, the key to this decision is the mate's likelihood of being impregnated by another male, either as a result of the current extra-pair copulation or as a result of future extra-pair copulations.

If the male perceived from his mate's actions that she had mated with other males under any number of circumstances, and that she was likely to continue to do so, his best evolutionary option was probably to abandon her. Indeed, the intensity of male sexual jealousy, in view of the associated risk that the female would leave the relationship, suggests that abandonment often may have been the best option even in an unambiguous case of rape. However, if the circumstances of the recent event indicated that the female was likely to mate with another male only after resisting to the best of her ability or if there was a reasonable likelihood that such resistance would have brought death or serious bodily harm upon the victim or her offspring (likely to be her mate's offspring), the evolutionarily sound option for the male often may have been to continue to invest in her. The reason the threat of death or serious bodily injury to the woman or those she commonly protects is expected to have influenced the mate's decision is that it was usually only under those circumstances that the reproductive interests of the victim's mate might be best served by the victim's submitting to the copulation rather than resisting to the best of her ability. If the copulation involved only threats of lesser consequences, the victim's mate's reproductive interests were probably best served by the victim's resisting the copulation and suffering those lesser consequences. Hence, males would have been selected to cease investing in females who copulated with other males under less coercive circumstances.

Natural selection, however, may have favored males who continued to invest in mates who, on clear evidence, had been raped. The need for clear evidence probably explains why males are so often suspicious of a mate's claims of having been raped. Selection for such male paranoia is implied in anthropologist Nancy Thornhill's statement that "mates of rape victims might be quite suspicious of alleged rape, preferring to view the sexual assault of their wives/girlfriends as simply adulterous liaisons" (1996, p. 93). "The doubt that mates of rape victims seem to exhibit about the victim's credibility," Thornhill continues, "is expected as a paternity protection mechanism." In view of this, selection would have favored females—and the kin of both victims and their mates—who made the same distinction between rape and other extra-pair copulation. We hypothesize that this is why a distinction has been made between the much larger category of sexually coerced copulations and the smaller category of copulations referred to by the word 'rape'.

Rape's threat to paternity certainty, the even greater threat to paternity certainty posed by consensual affairs, and the resulting selection for male suspiciousness about rape claims can account for many aspects of rape laws. For example, one manifestation of male suspiciousness about rape claims is concern about the victim's previous sexual conduct. As Spohn and Horney (1992, p. 25) observe, the "notion that the victim's prior sexual conduct was pertinent to whether or not she consented was based on the assumption that chastity was a character trait and that, therefore, an unchaste woman would be more likely to agree to intercourse than a woman without premarital or extramarital experiences." The legal scholar Susan Estrich (1987, p. 48) similarly observed that "in a general sense, the belief that a woman's sexual past is relevant to her complaint of rape reflects, as does the resistance requirement, the law's punitive celebration of female chastity and its unwillingness to protect women who lack its version of virtue." We suggest that the law's celebration of female chastity and its unwillingness to protect unchaste women reflect the human male's evolved preference to invest only in chaste mates.

Far from being a moral prescription about male behavior, this evolutionary analysis of why males and the rape laws they have formulated are so uncharitable toward victims' claims of rape may be of help to those seeking changes in the rape laws. Indeed, this help may already be starting to manifest itself.

In their writings and at law conferences, the legal scholars Owen Jones and Jack Beckstrom have been introducing lawyers and law professors to evolutionary work, including ours. They have argued that a better understanding of the psychological mechanisms influencing patterns of rape—an understanding informed by evolutionary biology—might make law more effective in deterring rape (Beckstrom 1993; Jones 1999). To the extent that knowledge about the causes of things becomes a part of the environment and increases our ability to change things, men who are made aware of the evolutionary reasons for their suspicions about their wives' or girlfriends' claims of rape should be in a better position to change their reactions to such claims. Legal reformers who were aware of the evolutionary reasons for the present laws could make more persuasive arguments for their reform, and their arguments might be better received if those to whom they were presented had basic knowledge of evolution. Further, the evolutionary understanding supports the view that rape laws can be changed by placing more women in the position to make and enforce such laws because women have evolved different attitudes toward rape.

The movement to reform rape laws might also benefit from a better and more widespread understanding of how and why deception is used by some women. Many legal scholars have interpreted the traditional requirement of corroboration in rape cases as indicating distrust of women (Spohn and Horney 1992). The distrust may reflect how both men and women think about women in regard to their allegations. Relative to men, women have evolved to avoid physical risks and physical harm more and to be less interested in status and dominance (Campbell 1995; Campbell et al. 1998; Geary 1998; Walston et al. 1998). The works just cited, especially those of the evolutionary psychologist Anne Campbell, also show that women have evolved to compete for limited resources and mates not so much by direct physical aggression as by indirect and low-cost (relative to physical aggression) means. In fact, research shows that in social competition human females use a sophisticated suite of indirect, low-cost tactics. Girls and women, relative to boys and men, tell more false stories about adversaries, gossip about them, start rumors about them, and use ostracism and manipulation of public opinion as tactics (Feshbach 1969; Brodzinsky et al. 1979; Cairns et al. 1989; Ahmad and Smith 1994; Bjorkqvist et al. 1994; Crick and Grotpeter 1995). We know of no studies of social knowledge that males and females differ in these ways, but we predict that such studies would reveal that such knowledge exists. Thus, the requirement for corroboration in rape cases may reflect, in part, evolved knowledge of the tactics females may use in social competition.

We suggest, also, that people are especially concerned about the credibility of women's allegations when sex is involved. As we have mentioned, people everywhere understand sex to be something that women have and that men want. This intuition about social life arises from the sex difference in minimum investment necessary for the production of offspring. That males want sex itself appears to have selected, in human evolutionary history, for females who used sex and promises of sex to manipulate men and get resources from them. Clearly, women behave this way far more often than men. Studies reveal that, relative to men, women seem to be more deceitful about their sexual interest in individuals of the opposite sex (e.g., behaving as if sexually interested when in fact they aren't), about sexual arousal (e.g., faking orgasm), and about personal sexual history (e.g., claiming to have had fewer partners than the actual number) (Buss 1994; Thornhill et al. 1995; Geary 1998). Studies also suggest that women are more deceitful with respect to mateship infidelity (Baker and Bellis 1995; Gangestad and Thornhill 1997b). Thus, especially when sex is involved (as it is in rape), there may be an evolved intuition that women sometimes lie for their own gain.

This is not to say that men don't lie about sexual matters. They obviously do, and presumably for personal gain, because a high number of sex partners is associated with high status and high self-esteem in men, and not in women (Quinsey and Lalumiére 1995). However, social intuition about women's use of sexual allegations, in combination with their use of low-cost competitive tactics, may lead to skepticism and to reluctance to judge in favor of a woman who "cries rape."

False rape allegations have received little systematic study. To some feminists, the concept of false rape allegation itself constitutes discriminatory harassment (Grano 1990). However, a careful study of 109 rape cases in the United States found 41 percent of rape accusations to be false as evidenced by the women's own recantations (Kanin 1994). The women studied gave three reasons for their false reports: providing an alibi for a consensual sexual encounter that might have led to pregnancy, seeking revenge against a rejecting consensual male partner, and obtaining sympathy and attention from kin and/or friends. Kanin emphasizes that false rape allegations "reflect desperate efforts to cope with personal and social stress situations" (p. 81).

One major goal of the movement to reform rape law has been to establish the view that rapists are motivated by a desire for violence rather than by a desire for sex. Indeed, in many states rape has been redefined as "sexual assault (or criminal sexual misconduct, etc.) to emphasize that rape was a violent crime and not a crime of uncontrollable sexual passion" (Berger et al. 1988, p. 331). Morris (1987, p. 177) argued against the notion that "rape is sexually motivated," and Miccio (1994, p. 82) told the Congressional Subcommittee on Crime and Criminal Justice that "the cultural myth that suggests that rape is a crime of passion . . . must be debunked." In a discussion of the legal implications of theories of rape, Fuller (1995, p. 159) lamented the fact that the assumption "that rape is a sexual act" is ever made in legal cases. Blatt (1992, p. 832) criticized international laws based on the assumption "that rape was a sexually, rather than politically, motivated offense." K. Baker, who at least acknowledged "that some rapes are predominantly about sex" (1997, p. 556), still claimed that for a significant number of rapists "the act of controlling—not sex, is critical to their motivation to rape" (ibid., p. 609).

In 1994, after hearing Eleanor Smeal of the Fund for a Feminist Majority testify that "rape is never an act of lust" (quoted in Shalit 1993, p. 7B), the US Congress passed the Violence Against Women Act, which created a new civil rights cause of action for "crimes of violence motivated by gender." "To be 'motivated by gender,'" Jones (1999, pp. 921–922) writes, "violent crimes must be: a) 'committed because of gender or on the basis of gender'; and b) 'due, at least in part, to animus based on the victim's gender.'" In light of the numerous flaws in the claim that rape isn't sexually motivated, such a legal emphasis on the notion that rape is a non-sexually-motivated "hate crime" is unfortunate. Law is influenced by assumptions about human nature (Jones 1999). Invalid assumptions negate law's credibility and potential for a more humane and just course of action. It is our position that legal issues pertaining to rape would benefit from knowledge of rape's causation and of evolved cognitive biases about rape victims.

Statutory Rape

Statutory rape is defined as sexual intercourse, consensual or not, with a female who is under the legal age of sexual consent. Understanding why the age of sexual consent is legislated requires an understanding of why rules, including laws, exist: They arose from past selection for the ability to control others in the social environment (Alexander 1979, 1987).

As was revealed by Richard Alexander's (1979) pioneering treatment of law in a Darwinian perspective, rules—whether or not codified in law—generally serve the interests of the powerful. This means that they serve the interests of adults more than those of children, the interests of men more than those of women, the interests of the rich more than those of the poor, and the interests of high-status males more than those of low-status males.

Females below a certain age are believed to lack mature judgment, and specifically to lack the ability to exercise adaptive mate choice. This is probably true of children and very young teenagers, so statutory-rape laws probably serve the interests of those individuals and their parents. Other factors, however, may be needed to account for the application of statutory-rape laws to older adolescents, who probably possess a sophisticated psychological adaptation for mate choice.

The application of statutory-rape laws to women as old as the late teens may have to do with their fathers' (and in some cases other genetic relatives') interests. Humans, like all other organisms, are evolved to pass on their genes by means of their offspring. Egg bearers are a limiting resource for the population's sperm bearers. An egg bearer's future parental investment is precious and should be expended only in the best circumstances for reproductive success. Parents may try to manipulate, even coerce, long pre-sexual periods for daughters, thereby making them more valuable on the mateship market because men prefer to invest their parental efforts in women with restricted sexual histories. Parents may also try to manipulate the romantic relationships and the mateships of their children, and especially their daughters.

In most societies, daughters have been viewed as their father's property, to be provided to certain others in exchange for alliance, assistance, or other resources. Many societies that lack codified laws have as a standard practice *bride price* (i.e., how much a man pays for a bride) or *bride serv-*

ice (how much work the man will do for his bride's family). According to the biblical scholar John Hartung (personal communication), the Codes of Maimonides and the earlier codes upon which they are based suggest that this is relevant to the concept of statutory rape. These codes defined statutory rape as sexual intercourse with a wealthy man's unmarried daughter, which required monetary compensation to her father. Also relevant are the fact that a daughter's value to her father was considered greatest when she was a virgin and the fact that an attractive daughter was considered more valuable than an unattractive one.

We suggest the following explanation for the expansion of the legal definition of statutory rape to include intercourse with females old enough to make informed mate choices: An individual female's reproductive value (ability to contribute offspring to the population in the future) is at its maximum just after she reaches puberty. As rated by people in general (not just men), this is also when a female's attractiveness is at its peak (Symons 1979, 1995; Johnston and Franklin 1993; Quinsey et al. 1993; Jones 1996). Her attractiveness makes her value as a mate maximal at this time. In preindustrial societies, women married at this age. The evolutionary anthropologist Elizabeth Cashdan (1996) proposes that women have a psychological adaptation for bonding to their chosen mate at this age. Cashdan views later mateship as less strongly bonding for the female because her lower mate value later in life is more likely to lead to shorter-term relationships. In the environment of evolutionary history, the woman's first mate choice often would have been the best she could do, because of her maximal attractiveness to men able and willing to invest in a mate. This is consistent with women's often-reported conscious view that a "first love" is different from, more significant than, and/or better than later romances (Cashdan 1996). Although parents may say and believe that they are trying to control a daughter's sexuality because of her lack of age-related wisdom, such control may actually be an effort by the parents to get the daughter past the time at which she is most likely to develop a strong romantic attachment on her own, so that they will have a better chance of controlling her eventual marriage.

This is not to exclude parental wisdom about how a daughter's sexual history influences her prospects for gaining an investing mate. Because of the evolution in humans of very long pre-adult life, which functionally is associated with the learning of cultural information that prepares an individual for reproductive success as an adult, older daughters are wiser about social life than younger ones. Thus, parents who effectively control pubescent and adolescent daughters' social lives may increase the daughters' mate value and their chances of obtaining wealthy, caring husbands. Fathers are especially controlling of the sexuality of daughters (Wilson and Daly 1981; Flinn 1988). Flinn's research in Trinidad has shown that a father's presence during a daughter's upbringing increases the probability that the daughter will marry a husband with significant wealth and land. Flinn found that daughter-guarding fathers actively (and sometimes violently) repelled their daughters' suitors.

Like any other component of rape law, statutory rape is based on evolved psychology. Central here is the motivation of parents (especially fathers) to limit the pre-marriage sexual behavior of their daughters. An understanding of this might help lawyers, judges, and juries to better serve the interests of all parties involved in a statutory-rape case. An evolutionary approach to law would focus on the often-conflicting interests of individuals and on how certain laws may reflect the interests of individuals other than the direct victim. Rape laws in general greatly reflect the interests of the victim's mate, and laws pertaining to statutory rape may reflect the interests of the victim's parents.

Punishment

The idea that punishment can influence the frequency of rape is far from unique to the evolutionary approach, as is evident from the fact that rape is punished in all known societies.[4] However, only the evolutionary approach asks why and to what degree certain environmental stimuli constitute punishment (Wright 1994).

Because psychological adaptations change in sex-specific ways over the course of an individual's life in correspondence to the environmental challenges our ancestors faced in their various life stages (Geary 1998), an individual is expected to perceive as punishments environmental conditions that were particularly severe obstacles to the reproductive success of our ancestors of the same age and the same sex as that individual. Since the majority of rapes are committed by males in their teens or their twenties, the punishments most effective in deterring rape may correspond to the obstacles faced by our male ancestors at those ages. Since it is during the teens and the twenties that competition for status and for sexual access to females is most intense and most crucial to a male's reproductive success, punishments that impair such competition may be the most effective deterrents to further rape. One measure that comes to mind in this regard is incarceration. Long incarceration at least partially removes the offender from the everyday male-male status pursuits that young men spend so much time practicing. Monetary penalties also come to mind; however, by increasing a rapist's disenfranchisement, such penalties may make him *more* likely to rape again.[5]

We do not propose a specific program for increasing the costs associated with rape; we simply suggest that social engineers who wish to get realistic about rape pursue a program of punishment that is informed by what is known about evolution.

"Chemical Castration"

Literal castration, which has been widely used as a costly punishment for rape in various countries, seems out of the question in latter-day Western societies. However, a debate over so-called chemical castration—the use of anti-androgen drugs in the treatment of rapists—is now underway in several US states. The social science explanation's claim that rape is caused by non-biological cultural forces and has nothing to do with sexual desire plays a central role in this debate. Indeed, a major objection to the use of such drugs—discussed in the literature reviewed by Willie and Beier (1989)—is the assertion that "it is not the sex hormones which represent the decisive driving force [in sex offenses], but psychological factors." Failure to understand that psychological factors *are* biological is also evident in Icenogle's (1994, p. 279) question as to "whether a state may elect to sentence convicted sexual offenders to the use of a *biological* treatment, as opposed to incarceration or *non-biological* psychotherapy." The mistaken notion that biology excludes environmental factors is evident in Tsang's (1995, p. 409) argument against the view that "it is biology that determines pedophilia, not the environment."

Once the true meaning of biology is grasped, the argument that anti-androgen drugs should not be used because they influence only the biological aspects of human behavior becomes absurd. Rape, like every other behavior of living things, is biological. Hence, any attempt to change this behavior will, by definition, involve influencing human biology. Changes in the social environment (e.g., psychotherapy, educational courses, imprisonment, counseling, public humiliation), changes in the levels of circulating hormones (e.g., "chemical castration"), and changes in the genetic makeup of an individual are all equally biological means of influencing behavior.

In view of the pervasiveness of the notion that rapists are not sexually motivated, it is also not surprising that "many experts say that castration will not work because rape is not a crime about sex, but rather a crime about power and violence" (Hicks 1993, p. 647). Estrich (1987, p. 82) argues against the use of chemical castration on the ground that the position that "convicted rapists should have a choice between castration and imprisonment" is "a choice which makes sense only if their crime . . . is understood as a problem . . . of uncontrollable sexual desire." Goldfarb (1984, p. 4) quotes Nicholas Groth as stating that offering rapists the option of chemical castration "reflects the misconception that most sex offenders are raping out of some type of sexual desire." Vachss (1994, p. 112) argues that "such a 'remedy' [as chemical castration] ignores reality," in that "sexual violence is not sex gone too far; it is violence with sex as its instrument." Tsang (1995, p. 400) asserts that "drug therapy for rape cases . . . goes against the feminist view of rape as a crime involving violence and domination of women, and assumes rape to be primarily a sexual act," and that "feminists would argue that reducing the 'libido' would do nothing to reduce the threat of violence." Spalding (1998, pp. 132–133) even asserts that "because [rapists] are motivated not by sexual drive, but by intense feelings of hatred and hostility, the procedure [of chemical castration] may cause an increase in the occurrences of this type of sexual battery."

Although some of the influences of gonadal hormones on male sexual motivation have been known for decades, most social scientists have ignored or rejected the possibility of using hormones to reduce the frequency of rape because of their general view that anything "biological" is irrelevant to a "culturally" determined behavior and because of their specific adherence to the dogma that rapists are not sexually motivated (Cohen et al. 1971; MacDonald 1971; LeGrand 1973; Rada 1978a; Groth 1979; Katz and Mazur 1979; Dusek 1984). Hence, proponents of the social science theory of rape assert that, at best, hormonal treatments might cause potential rapists to switch from rape to some other form of violent aggression against women. Although the evidence on the effect of hormonal treatment is limited, there is little or no evidence to support the assertion that either castration or hormonal treatment leads to non-sexual aggression, and there is considerable evidence to suggest that they reduce sexual crimes (Kopp 1938; Bremer 1959; Sturup 1960, 1968; MacDonald 1971; Rada 1978b). Any decision on whether to use such drugs should be based on how they actually affect behavior.

8

Social Influences on Male Sexuality

Evolutionary psychologists, contrary to common expectation, subscribe to a cardinal doctrine of twentieth-century psychology and psychiatry: the potency of early social environment in shaping the adult mind.
—Robert Wright, *The Moral Animal* (1994), p. 8

The evolutionary approach has more to say about the early social environment than some of its critics may think. As Wright (1994) says, "if we want to know, say, how levels of ambition or of insecurity get adjusted by early experience, we must first ask why natural selection made them adjustable." The same is true of levels of sexual restraint and of willingness to use violence to obtain desired goals. Although some individual differences in these behaviors may be due to genetic differences (Ellis 1989), "a larger role is played by genetic commonalties: by a generic species-wide developmental program that absorbs information from the social environment and adjusts the maturing mind accordingly" (Wright 1994). Even when a behavior is heritable, an individual's behavior is still a product of development, and thus it has a causal environmental component.

An example involving heritable resistance to an infectious disease should be illustrative. An individual who is genetically predisposed to infection with the disease cannot get it without encountering the infectious agent. And the genes in question may not affect a disease state, even when the individual encounters the agent, if the individual is well nourished and capable of combating the agent.

The evolutionary model views the human brain as a bundle of numerous specialized adaptations created by specific, evolved gene-environment interactions during their ontogeny. After their ontogenetic construction, these adaptations interact with specific aspects of the environment to produce rape.

Essentially all men have sexual psychological adaptations designed for obtaining a large number of mates. However, heritable adjustments in the details of certain sexual adaptations in response to environmental cues processed during development probably create some individual differences in ease of activation of these adaptations. The mechanisms that make such adjustments are facultative—that is, dependent on specific environmental variables. Even if there are significant genetic differences among individual men in some or all of the psychological adaptations that underlie rape, to fully understand rape and to reduce it we will have to determine how environmental differences affect the propensity to rape. The same holds if the psychological adaptations that generate rape reflect multiple sexual adaptations that exist in a mix in the population of men as a result of frequency-dependent selection (as may be true of psychopathic versus non-psychopathic phenotypes, for example).

It is important to realize that "this emphasis on psychological development doesn't leave us back where social scientists were twenty-five years ago, attributing everything they saw to often unspecified 'environmental forces'" (Wright 1994, p. 82). Instead, an understanding of the ultimate evolutionary reasons why humans have facultative adaptations that respond to variables in the social environment greatly enhances our ability to specify what social variables influence development in what ways. This also gives the evolutionary approach an advantage over approaches that use arbitrarily chosen environmental factors to explain rape. For example, many proponents of the social science theory of rape (e.g., Denmark and Friedman 1985; Stock 1991) hold that one specific way in which males in some cultures are taught to rape is through the viewing of violent pornography, which inspires imitative behavior. The social scientists pushing this notion, however, cannot explain why the human brain is purportedly structured so as to respond in this specific way to the specific environmental stimulus of violent pornography. Why, for example, should males seek out and imitate violent pornography but not other human activities depicted in videos? There is no consideration of the ultimate basis for the asserted proximate explanation, no sound theoretical foundation for it. Aside from the obvious fact that violent pornography cannot account for the historical and cross-cultural (indeed cross-species) occurrence of rape, such an arbitrary environmental explanation is refuted by everything we know about biases in human development, perception, cognition, emotions, and motivation. It also has a logical flaw: An environmental factor is identified as a cause of a human behavior without any attempt to explain why other kinds of environmental variables that could conceivably also influence the same category of behavior do not do so. Consequently, although the viewing of violent pornography may figure in the proximate causation of the raping behavior of some men, this view is severely limited in its ability to predict anything useful about rape or related behaviors. It cannot explain the data on who is raped, or the data on when and where rape occurs. Although the removal of violent pornography may be desirable in its own right, it is very unlikely to solve the problem of rape.

Rape occurs among humans under a wide range of "physical" and "cultural" environments—indeed, it occurs in all the environments in which humans societies have been known to exist. Hence, cross-cultural evidence actually indicates that a relatively narrow set of environmental changes (including the punishments mentioned in the previous chapter and the structural environmental barriers described below) might be needed to reduce the incidence of rape significantly. The real lesson to be drawn from cross-cultural studies is *not* that rape will vanish with the end of patriarchy.

Once the scientifically false beliefs that arbitrary learning is all-important in creating human behavior and that rape can be prevented simply by refraining from teaching males to rape are abandoned, they can be replaced with ideas derived from the evolutionary model. Those ideas can then provide direction for efforts to prevent rape by changing the identified aspects of the environment.

We agree with social scientists that males should be educated not to use force or the threat of force to obtain sex. However, we suggest that educational programs aimed at preventing rape would be much more successful if they would focus on the goal that motivates males to use such tactics. In direct contrast to the social science explanation of rape, the clearest implication of evolutionary theory is that the motivation for rape is a result of the differences between male and female *sexuality*. That is, *the evolved psychological adaptations that produce male sexual motivation are necessary proximate causes of rape*. It follows that creating environmental conditions that will decrease the frequency of rape requires identification of the exact nature of the psychological mechanisms that guide male sexual behavior. The more we understand how these mechanisms develop and what cues they respond to, the better we will be able to modify male sexual development and associated male sexual behavior.

Not only are ultimate explanations of male sexual motivation ignored by many of those who wish to prevent rape; the potential importance of such knowledge is actively *denied* by most social scientists and by nearly all academic feminists. For these individuals, the key to preventing rape is convincing men and women that rape is a political act that has nothing to do with biological differences between male and female sexuality. Indeed, the idea that rapists are sexually motivated is often considered to be a "rape myth"[1] that must be eradicated by "education." For example, Fonow et al. (1992, pp. 118–119) suggest that "feminist rape education needs to address the themes of rape as sex and rape as social control" and report that "women's rejection of rape as sex was reinforced and supported through the education; men's beliefs were confronted, but perhaps not forcefully enough." Syzmanski et al. (1993, pp. 54–55) claim that their "rape awareness workshop appears to have been an effective educational forum" insofar as "subjects who had not attended an awareness workshop . . . thought that sex was a motivation for rape . . . significantly more than did those who had attended the workshop." According to Stock (1991, p. 73), "for sexual coercion to cease, women must accrue enough power through increased access to concrete resources, expertise, and status to make it less possible for males to continue to maintain constructs and beliefs that stipulate male domination of females." Repetition of the claim that rape is not motivated by sexual desire is the greatest obstacle to the creation of more effective means of preventing rape.

Rejection of the overwhelming evidence that rapists are sexually motivated appears to be grounded in the belief that rapists driven by lust might not be considered responsible for their actions. Yet, to our knowledge, proponents of the social science model never assert that the supposed male drive to control and dominate women excuses a rapist's behavior, even though that alleged motivation is claimed to be so powerful as to account for nearly all aspects of male-female relations. (Again, scientific explanations of behavior only provide information about the causes of actions;

they imply nothing about who should or should not be held responsible for their actions.)

In reality, the role of sexual motivation as a cause of rape should be a reason for optimism about future attempts to prevent rape. Consider these two points:

- Many men don't rape and are not sexually aroused by laboratory depictions of rape. This suggests that there are cues in the development environments of many men that prohibit raping behavior.
- Acknowledging the role of sexual motivation allows the formidable and rapidly increasing body of scientific knowledge about the evolution of male sexuality to be applied to identifying the cues that prevent rape and to lowering the frequency of rape.

The first step in understanding how the "social environment" (that is, the behavior of other people) influences the ontogeny of male sexuality is to remember the crucial finding from the evidence on non-human species: *male sexual pursuit of unwilling females commonly emerges from ontogenies that lack any sexual socialization.* That is, rape occurs even when males are not encouraged to rape. In view of this fundamental aspect of male sexuality, it is not surprising that among humans, contrary to much social science writing on gender roles, "the great majority of prescriptive messages concerning [male sexuality, including rape] are intended to suppress it, not to foster it" (Symons 1979, p. 303). Indeed, any explanation of the species-typical sexual behavior of human males must be able to account for the universal presence of "moral traditions" that "limit male [sexual] activities" (ibid., p. 246) through "a learned tendency to avoid performing sexual acts under certain conditions" (LeVine 1977, p. 222).

The ethnographic record indicates that rape is universally discouraged by moral traditions. In all cultures, rape of at least some women under at least some circumstances is considered immoral (Palmer 1989a). An evolutionary understanding of why traditions limiting rape have come to exist may provide clues to the proximate mechanisms by which they operate and, hence, to ways of more effectively limiting rape in modern populations.

We suggest that the interaction of genes with environmental factors that included social influences restraining certain forms of sexual behavior resulted in a type of male sexual flexibility that was evolutionarily advantageous in human ancestral settings. Specifically, it increased the ability of males to avoid extremely costly forms of sexual behavior (and rape is, under many circumstances, extremely costly) and to still be motivated to seek out and take advantage of less costly sexual opportunities. A male whose sexuality developed in the absence of the restraining influence would be more likely to engage in sexual acts whose potential costs (injury or death) greatly outweighed their reproductive benefits, such as attempting to rape a woman when her husband or father was nearby. On the other hand, males influenced by their relatives to refrain from too great a range of sexual activities would also be at an evolutionary disadvantage. This accounts for the universal presence of traditions condemning rape, but often only under certain circumstances. Because rape is condoned or even encouraged when the victims are members of an enemy group, Murphey (1992, p. 21) can correctly point out that moral traditions cause "males to abhor rape"; by the same token, feminists can correctly stress that the condemnation of rape is inconsistent, and that the degree of outrage over rape corresponds more with the interest of males than with that of females (Clark and Lewis 1977; Dietz 1978).

Exactly how the socialization of boys figures in the proximate causes of men's propensity to rape or to refrain from raping is not completely known. However, evolutionarily informed researchers have gathered considerable knowledge about the proximate mechanisms involved in how children are inculcated. A comprehensive cross-cultural analysis of the training of boys and girls published by the biologist Bobbi Low (1989) illustrates important universalities that reflect the evolved sex differences in human sexual psychology. Beyond the general worldwide differences in how girls and boys are educated, Low's study shows that, across societies, the more polygynous the society (i.e., the higher the potential offspring production for males via multiple wives) the more often sons are taught specific ways to compete socially. Low also found that the more stratified the society the greater is the emphasis on girls' being sexually restrained and obedient and the less likely girls are to be urged to be self-reliant.

Low's study was based on her knowledge of the role of evolution by selection in shaping the psychological mechanisms that guide learning and teaching. It was inspired by the hypothesis that sons and daughters will be trained and will learn differently in ways related to the respective evolutionary histories of reproductive success of the two sexes in a polygynous social environment. In general, sons are taught how to become polygynists, and daughters are taught how to deal with proprietary male relatives and prospective husbands. Sexual restraint and obedience in women are desirable to men because men desire to invest in offspring they have sired. The teaching of sexual restraint and obedience to females is most exaggerated in stratified societies because hypergyny (females' marrying up the social ladder) is common in such societies and thus families compete to place daughters as mates of high-status men (Dickemann 1979a,b, 1981). Low's research makes it clear that sexual socialization is not an arbitrary cultural practice but, rather, reflects psychological adaptation in adults for teaching social skills to children and sexually dimorphic psychological adaptation in children for learning social skills.

The cross-cultural findings on how boys and girls are socialized are not evidence that boys are taught to rape. Instead, they suggest why social disapproval of rape may vary in intensity. Consider, for example, the evidence that men with high rape proneness have higher incidences of reduced investment by parents (e.g., absence of the father) and of negative or unproductive heterosexual interactions (Malamuth and Heilmann 1998). Malamuth and Heilmann argue that these circumstances provide developmental cues that male minds have evolved to track because those cues would have provided information about the relative likelihood of mating success with non-coercive versus coercive sexuality in human evolutionary history.[2] This approach appropriately considers the development of men's sexuality as adaptation and as therefore subject to specialized socialization experiences that would have affected reproduction in the environments of deep human history.

In the United States, adolescent, high school, and college males, when presented with sexual scenarios, have been found to be less likely than same-age females to interpret the sex as forced and are more likely to blame the victim (Lonsway and Fitzgerald 1994; Cowan and Campbell 1995). Although sex-specific social learning may be one proximate cause of this male attitude, other experiences that are involved in the ontogeny of men's sexual psyches may also account for these findings.

Males' tendency to hope that females are sexually interested and perhaps even sexually available even when they really aren't promotes males' continual pursuit of mates. Evolutionary theory predicts that, during ontogeny, some male sexual adaptation constructs an incorrect understanding of female sexuality that promotes the adaptive goal of many mates. In some environments, such an adaptation may filter out much social encouragement of sexual restraint and respect for women's wishes, merely allowing the incorporation of the social messages that lead to avoidance of the gravest social costs of sexually accosting many women (e.g., being killed by a jealous husband).

The modern understanding of how phenotypes are inherited through the replication of both genetic and environmental conditions suggests that male cultural traditions—behaviors copied by sons from their fathers—are likely to be crucial in creating socialized inhibitions against committing rape. That is, the presence of a father (or a father substitute) exhibiting restrained and non-exploitative sexual behavior may be a critical factor in the ontogeny of restrained sexual behavior in young human males. Evidence of the effect of a father's absence on a son's sexually coercive behavior (and on criminal behavior in general) supports the importance of the role of the father in this socialization (MacDonald 1988, 1992; Surbey 1990; Lykken 1995; Barber 1998; Malamuth and Heilmann 1998). Nigel Barber's (1998) review of the literature on the relationship between absence of a father (and associated marital instability) and criminal activity suggests that children raised without the presence of a father are about seven times as likely to be involved in serious criminal behavior throughout life. Although a mother's parental investment in her children is typically more "hands-on" than a father's (Geary 1998), Barber (1998, pp. 5–6) points out that "children growing up in a father-present home learn to interpret the world very differently because they are exposed to masculine perspectives as well as feminine ones." "This," Barber continues, "transforms a child's view in respect to issues of social stability of the household, the degree of kindness to be expected from unknown others, the nature of the relationships between the sexes, the value of work and accomplishment, and the necessity to follow social rules."

The claim that male cultural traditions can deter rape is in direct opposition to the feminist assumption that male cultural traditions cause rape, and the two views lead to fundamentally different solutions. The feminist view predicts that rape can be prevented only by a wholesale abandonment of male traditions—that rape "is not an isolated act that can be rooted out from patriarchy without ending patriarchy itself" (Griffin 1971, p. 36). This view would seem to suggest that boys would be better off without paternal presence. In reality, though many aspects of patriarchal traditions may be undesirable for a variety of reasons, the abandonment of all male traditions that might be deemed patriarchal would be likely to *increase* the frequency of rape. Lederer (1980, p. 124) quotes the sociologist Judith Bat-Ada as having observed that there are many dangers inherent in removing males "from the traditional view of male-female, father-child relationships, which, although patriarchal, at least involved some norm of responsibility and concern."

On the other hand, there is no reason to assume that moral traditions correspond perfectly to what might be considered desirable behavior. Moral traditions are based in psychological adaptations, and their function lies in enhancing the status, the survival, and hence the reproductive success of individuals who create and perpetuate them. Thus, these traditions are not necessarily behavior that corresponds to what *everyone* considers "good" (Alexander 1987). Moral traditions may be inconsistent with regard to rape; for example, rape in war is endorsed in the Bible (Hartung 1992). Hence, rape cannot be reduced either by abandoning all traditions or by simply adhering to traditional attitudes.

Far from opposing the social science theory's assertion that changes in culture (i.e., in the behavior of people in their environment) are crucial to lowering the frequency of rape, the evolutionary approach identifies specific cultural changes that are likely to be much more effective in preventing rape than those proposed by adherents of the social science theory. Just as important, the evolutionary approach provides strong evidence that some of the cultural changes proposed by adherents of the social science theory might well make rape more frequent.

9

Educational Programs

The correlation between the age distribution of rapists and the peaking of the male sexual drive in the teens and the twenties suggests that steps to prevent males from raping will be most effective if they focus on males during and before this age range, but only if they are really educated about their sexuality. Although rape-prevention methods based on the social science model also often stress the importance of influencing young males, they do not focus on male sexual impulses. For example, Parrot (1991, p. 131) states that the most crucial thing to teach boys is that rape "is a crime of violence motivated by the desire to control and dominate." In essence, such "education" tells boys that, as long as their acts are motivated by sexual desire, they cannot be committing rape.

We envision an evolutionarily informed educational program for young men that focuses on increasing their ability to restrain their sexual behavior. Completion of such a course might be required, say, before a young man is granted a driver's license.

Such a program might start by getting the young men to acknowledge the power of their sexual impulses and then explaining why human males have evolved to be that way. A good starting point would be the evolutionary reasons why a young man can get an erection just by looking at a photo of a naked woman, why he may be tempted to demand sex even if he knows that his date truly doesn't want it, and why he may mistake a woman's friendly comment or her tight blouse as an invitation to have sex when in fact sex is practically the last thing on her mind. After each of these points, it should be emphasized that the reason a young man should know these things is so he can be on guard against certain effects of past Darwinian selection. The fallacy of the naturalistic fallacy—that a young man's evolved sexual desires offer him no excuse whatsoever for raping a woman, or for harming the interests of another person in any other way—should also be strenuously emphasized. Most important, the program should stress that, if he understands and adamantly resists his evolved desires, a young man may be able to prevent their manifestation in sexually coercive behavior. We suggest that the program conclude with a detailed and graphic discussion of the penalties for rape, including how much time a convicted rapist is likely to spend in prison and what conditions he can expect to encounter there. Though hypothetical at present, such a course may become a real possibility once the evolutionary basis of rape is widely understood.

A program of anti-rape education for females should begin with the same explanation of *male* sexual adaptations that should be used in the program for males. In addition to that and some instruction in self-defense, we suggest that the program address several matters that are typically ignored or denied by the social science model. As Mynatt and Allgeier (1990, p. 121) point out, "the identification of characteristics that are associated with high levels of risk for sexual coercion has received little attention. . . . However, educational programs aimed at reducing the vulnerability of women to sexual coercion are dependent on the acquisition of information concerning risk factors. . . ."

Contrary to the social science explanation's claim that the sexual attractiveness of the victim has no effect on the rapist's motivation, there are certainly aspects of behavior and appearance that influence a woman's likelihood of becoming a rape victim.

The social science model not only denies that sexual attractiveness influences rapists; it also holds that the selection of a victim is determined—perhaps solely—by her vulnerability. In reality, however, age is universally a powerful determinant of a female's sexual attractiveness, and sexual attractiveness influences the chances of being raped. A woman is considered most attractive when her reproductive value and her fertility are at their peak (i.e., from the mid teens through the twenties). Hence, the evolutionary approach predicts that tactics that focus on protecting women of these ages will be the tactics most effective in reducing the overall frequency of rape. And, in fact, this prediction is supported by the correlation between the age distribution of rape victims and the age distribution of female sexual attractiveness (Mynatt and Allgeier 1990). The educational program for young women should also address how other elements of attractiveness (including health, symmetry, and hormone markers such as waist size), and clothing and makeup that enhance them, may influence the likelihood of rape (Singh 1993; Grammer and Thornhill 1994). This is not to say that young women should constantly attempt to look ill and infertile; it is simply to say that they should be made aware of the costs associated with attractiveness.

Young women should also be informed that female choice, over the course of the evolution of human sexuality, has produced men who will be quickly aroused by signals of a female's willingness to grant sexual access (Buss and Schmidt 1993; Grammer 1993). Furthermore, women need to realize that, because selection favored males who had many mates, men tend to read signals of acceptance into a female's actions even when no such signals are intended (Buss 1994; Mynatt and Allgeier 1990).

And it should be made clear that, although sexy clothing and promises of sexual access may be means of attracting desired males (Cashdan 1993), they may also attract undesired ones. Women's dress is receiving considerable attention from evolutionarily informed researchers. Cashdan (1993) found that, relative to college women in environments the women perceived to be richer in potential investors in offspring, college women who perceived that the men in their social environments were *not* potential investors dressed "sexier" and were more likely to use sex as a tool for getting and keeping mates—that is, the women in the apparently more investor-rich environments were more conservative in dress and in sexual behavior. This conservatism is a female tactic to increase the confidence of paternity of men capable of investment and thereby secure their investment.

The evolutionary psychologist Nigel Barber (1998), who has examined dress length and other factors affecting skin exposure in women's fashions in the West, finds that, in general, women's dress follows patterns that can be predicted on the basis of whether sexual competition is more favorable for women or for men. When men outnumber women and have sufficient wealth to invest in women, styles of dress that depict sexual inaccessibility are most popular with women; when women outnumber men, the most popular fashions are less conservative. The ethologist Karl Grammer (1993), in a study of college women and women at bars, found that women at the midpoint of the menstrual cycle exposed the most skin. Although there is much more to be learned about women's dress behavior, these studies indicate that it is not arbitrary; rather, it is tactical, and it reflects adaptation for using clothing as sexual strategy.

Most discussions of female appearance in the context of rape have asserted that a victim's dress and behavior should affect the degree of punishment a rapist receives. These unjustified assertions may have led to the contrary assertions that dress and behavior have little or no influence on a woman's chances of being raped, not because there is convincing evidence that they don't, but out of a desire to avoid seeming to excuse the behavior of rapists to any extent. In one such counter-assertion, Sterling (1995, p. 119) writes that Amir's (1971) finding that 82 percent of rapes were at least partially planned indicates that "in most cases a woman's behavior has little, if anything, to do with the rape." The logic of Sterling's argument is questionable; it implies that behavior and appearance also have little if anything to do with being asked out on a date, since a date is usually planned. But, more important, Sterling's argument suggests that young women need not consider how their dress and their behavior may affect the likelihood that they will be raped. The failure to distinguish between statements about causes and statements about responsibility has the consequence of suppressing knowledge about how to avoid dangerous situations. As Murphey (1992, p. 22) points out, the statement that no woman's behavior gives a man the right to rape does not mean that women should be encouraged to place themselves in dangerous situations.

An informed attitude toward risk factors in rape might be promoted at universities, where women currently receive a very different "education." In women's studies courses, in other social science courses, and in "rape prevention handbooks," they are told that rape is not sexually motivated and not related to the above-mentioned risk factors. For example, the "Myth vs. Fact" section of a handbook currently used in the Rape Prevention Education Program of the Police Department of the University of California at Davis begins with these assertions:

MYTH—Sexual assault is caused by uncontrollable sex drives.

FACT—Sexual assault is an act of physical and emotional violence, not of sexual gratification. Rapists assault to dominate, humiliate, control, degrade, terrify, and violate. Studies show that power and anger are the primary motivating factors.

MYTH—Women provoke sexual assault, and sex appeal is of prime importance in selecting targets.

FACT—Sexual assault victims range in age from infants to the elderly. Appearance and attractiveness are not relevant. A rapist assaults someone who is accessible and vulnerable.

This politically motivated stance denies that men (non-rapists *and* rapists) have evolved sexual preferences for young and healthy women and are attracted to women who signal potential sexual availability by means of dress and behavior. It is dangerous to women because it misinforms them about male behavior. If young women really understood the evolved nature of male sexuality, they surely would be in a better position to avoid rape.

We endorse the common-sense view of rape proposed by Camille Paglia (1992, 1994). Paglia, who sees rape as sexually motivated, urges women to be skeptical toward the feminist "party line" on the subject, to become better informed about risk factors, and to use the information to lower their risk of rape. Evolutionary biology, Paglia notes (1996, p. 69), "is forcing science back onto the feminist agenda, where it has been disgracefully absent." Knowledge *is* power.

An educational program for young women might also address the likelihood that it was the absence of evolutionary theory in Sigmund Freud's thinking about the mind's structure that led to the widespread adoption of the myth that women subconsciously desire to be raped. (See Freud 1933.) That myth was widely accepted in law and medicine from the 1930s to the early 1970s (Kanin 1994). In reality, any desire to be raped must always have been selected against in human evolutionary history, since it would have interfered with the fundamental reproductive strategy of females—i.e., to choose mates on the basis of the benefits they are likely to provide.

10

Barriers

The anthropologist Robert LeVine (1977, p. 222) uses the term *structural barrier* to describe "a physical or social arrangement in the contemporary environment of the individual [male] which prevents him from obtaining the sexual objective he seeks." This suggests that people in general should be educated in how not to create situations in which women are especially likely to be raped. The goal is to increase the effort and/or the risk (including the risk of punishment) associated with rape to the point where rape becomes unlikely.

The idea of *physical* barriers (e.g., seclusion of women, still practiced in some cultures) is understandably abhorrent to many people. However, there are sensible and much less oppressive arrangements that might well lower the frequency of rape, such as locating a summer camp for teenage boys at the opposite end of a lake from a camp for teenage girls.

Social barriers simply involve the presence of other people in situations where males and females come into contact. Such barriers are present in most cultures, most often in the form of traditional patterns of movement that keep females—especially at the ages when they are most sexually attractive—out of isolated areas. Other social barriers include patterns of movement that keep males and females separated in certain circumstances (showers, bathrooms, sleeping quarters) and practices (such as chaperoning) that keep males and females from being isolated together. It is particularly important to keep males and females from encountering each other in isolated places where the costs of rape to the male are lowered. Among the Hewa of New Guinea, their ethnographer, the evolutionary anthropologist Lyle Steadman, noted in a personal communication, "men and women both assume that if a young woman is encountered in an isolated area by a man who is not closely related, that man will rape her."

An appreciation of the importance of structural barriers suggests many causes for concern in modern Western countries, where the common practice of unsupervised dating in isolated environments such as automobiles and houses, often accompanied by alcohol consumption, has placed young women in environments conducive to rape to an extent probably unparalleled in history. Any educational program dedicated to preventing rape should inform young women about these risks. Although it might be argued that reinstating structural barriers entails losses in personal freedom, the consequences of the absence of such barriers should also be considered.

The value of barriers is highlighted by the problematic nature of resistance. Self-defense training for women is a worthy part of rape-prevention education.[1] However, relying solely on defensive tactics entails potential risks that would be minimized if other preventive measures were taken too and if appropriate barriers were in place. If there is resistance, a rape attempt must already be underway. Not only is there no guarantee that resistance will prevent a rape; in certain circumstances, resistance may increase the woman's risk of greater physical injury or death. And it should be kept in mind that the degree of resistance needed to deter rape will vary with the victim's age and attractiveness.

Of course there are sensible measures that individuals can take, whether or not barriers are present. The sex researchers Elizabeth and Albert Allgeier (1991), upon examining the data on rape risk factors, advocate that men and women interact only in public places during the early stages of relationships, or at least that women exert somewhat greater control over the circumstances in which they consent to be alone with men.

11

Treatment and Recovery

The evolutionary perspective holds great potential for improving treatment of the post-rape psychological pain of rape victims and of their significant others. However, many guidebooks for rape crisis centers are based on the ideological assertion that "force, rather than sexuality, is the overriding feature of rape" (Morrison 1986, pp. 11–12). The notion that "one of the most misleading assumptions generally made with regard to men who rape is that their offenses are motivated by sexual desire" (Groth and Birnbaum 1986, p. 17) has remained an integral part of the strategies used by rape crisis centers (Scott 1993). In addition to replacing this political rhetoric with scientific knowledge about the proximate causes of rape, evolutionary theory can provide an ultimate perspective on human suffering and its alleviation.

As we have emphasized, evolutionary biology cannot provide moral guidance; however, insofar as moral values involve the reduction of human suffering, knowledge of what counts as human suffering is essential. Evolutionary biology explains human suffering in ultimate terms, as the outcome of the kinds of events that reduced the reproductive success of individuals in the past. Thus, evolutionary biology provides an objective way to define categories of human suffering. It also provides direction for studies of the cues that activate suffering. When these cues are fully understood, humans will have knowledge that is essential for reducing pain.

The suffering caused by a rape is not limited to the victim; it may extend to anyone who has an emotional attachment to her (e.g., a husband, children, other relatives). In evolutionary terms, emotional attachment occurs when there is an overlap in reproductive interests, as between a woman and her mate and her genetic relatives. The evolutionary prediction is that, all else being equal, the more a woman's reproductive success would have contributed to the genetic success of her mate or her relatives in evolutionary history, the greater the suffering of those individuals is likely to be after she is raped.

In an evolutionarily informed post-rape counseling and treatment program for victims and their significant others, the counselors would understand the proximate causes of the psychological pain and would direct the victims and their significant others to the sources of that pain. Psychotropic medications might be employed (selectively and cautiously, so as not to eliminate the defense that psychological pain provides). The victim's age, the change in her value to her mate and her family, the credibility of her rape report, and the paternity concerns of her mate would be taken into account. By acknowledging what is unique about the psychological pain surrounding rape and by addressing the anticipated magnitude of the pain, the evolutionary approach can focus therapy where it is most needed. Finally, such a program would inform the victim about cues that may increase the probability of rape, thus helping her avoid being raped again.

Whether evolutionarily informed therapy programs would alleviate the psychological pain generated by rape is an empirical question, as is who might benefit most from such programs. We hope, nonetheless, that such programs will be forthcoming. A therapy program that maintains that men rape because they collectively want to dominate women will not help a victim to understand why her attacker appeared to be sexually motivated, why her husband or boyfriend may view the attack as an instance of infidelity, why she can no longer concentrate enough to conduct her routine life effectively, or why her father wants to keep the attack secret. Programs based on ignorance of what is creating the victim's post-rape problems seem as useless as those based on the Freudian psychodynamic theory of rape.

We have mentioned the Freudian proposal that women desire to be raped. In addition, Freudian theory proposes that young human males wish to copulate with their mothers, and that, when this desire is unresolved, it produces sexual deviance in male adulthood, female rape victims being substituted for the mother (Freud 1933).[1] Evolutionarily informed people realize that pursuit of rape by women and incestuous psychology in men such as Freud posited cannot possibly have evolved. Those trying to help rape victims and their significant others need to have reality on their side.

12

Conclusion

The reason why the friend we mentioned in chapter 1 asked us about the causes of rape was that she was suffering. She hoped that our answer to her question would reduce her pain, would help her to avoid a recurrence of the event that caused the pain, and would help to prevent others from experiencing similar pain. The answers she had been given previously—that sex had been irrelevant to the act, that the man had been motivated by desires to control and dominate her, that a patriarchal culture had given the man these desires through childhood socialization, that rape was a means by which all men controlled the lives of all women in order to maintain the patriarchal culture, and that her appearance was not a factor in her chances of being raped again—were based on the social science explanation, and they seemed to fail to account for several things. Why, if he had not been sexually motivated, had the man used tender compliments in his attempts to initiate sexual acts throughout the evening? Why, after the rape, had he apologized for having resorted to physical restraint and to threats of further force? Why, if her appearance was not relevant to her chances of being raped again, was she now reluctant to dress as attractively as she had in the past?

The choice between the social science explanation's answers and the evolutionarily informed answers provided in this book is essentially a choice between ideology and knowledge. An evolutionary approach to rape provides the following answers to the questions we posed in the first chapter.

Why are males the rapists and females (usually) the victims?

This question can be answered at each of the two complementary levels of causation in biology: the ultimate and the proximate. We will begin with the ultimate, which is the more general and encompassing level of causation.

Males and females, both juvenile and adult, faced many sex-specific obstacles to reproductive success during human evolutionary history. As a result, selection favored different adaptations in the two sexes. Sexual selection—the primary kind of selection that explains the sex differences that lead to rape—is the differential reproductive success of individuals due to their trait differences that affect mating success (measured by mates' survival, parental investment, and reproductive capacity, and, in males, also by number of mates and by successful fertilization of eggs in competition with the sperm of other males). Sexual selection's action on each sex is governed by the relative parental investment of the sexes. Parental investment consists of the parental materials and services that determine the number and the survival of offspring; thus, it is the commodity for which each sex competes in the competition for mates. In humans, the parental investments of the two sexes may sometimes be nearly equal; however, the minimum parental investment for offspring production by a male is trivial: a few minutes of mating and the small amount of energy needed to place an ejaculate in a female's reproductive tract. The female must invest all the time and energy required for gestation, birth, and lactation.

The sex difference in the minimum parental investment is the key to understanding the sex-specific historical selection that gave rise to rape. Given the small investment of males and thus the low cost of each male mating, sexual selection favored males who achieved high mate number. As a result, men show greater interest than women in variety of sex partners and in casual sex without investment or commitment. Selection on females favored careful mate choice that allowed them to expend their precious parental investment under the circumstances most conducive to the production of viable offspring. Female adaptations for mate selection fall into two categories: (1) preference for males with status and resources, which evolved because such males provided material benefits to females and their offspring, and (2) preferences for males with physical markers in behavior and body of genetic benefits, which evolved because they increased the survival chances of a female's offspring.

Human rape arises from men's evolved machinery for obtaining a high number of mates in an environment where females choose mates. If men pursued mating only within committed relationships, or if women did not discriminate among potential mates, there would be no rape. The two leading evolutionary hypotheses for the existence of human rape behavior are (1) that rape is an incidental effect (a by-product) of men's adaptation for pursuit of casual sex with multiple partners and (2) that rape is an adaptation in and of itself. According to the first hypothesis, rape was indirectly sexually selected. According to the second, rape was directly selected because rape itself promoted success in competition for mates. Mutation-selection balance, drift, and other ultimate causes other than selection are not consistent with the data on rape's common occurrence and its high cost to rapists. Data also disconfirm general explanations of rape that are based solely on evolutionarily novel environments.

Ultimate explanations pertain to evolutionary agents that account for the existence of biological traits. Distinguishing the two leading ultimate hypotheses for rape empirically would require additional research along the lines discussed in chapter 3. Existing data do not allow a strong conclusion one way or another. It is entirely clear, however, that rape is centered in men's evolved *sexuality*.

The proximate causes of rape include genes, environmental cues, ontogeny, learning, physiology, and psychological and behavioral responses to environmental stimuli. The importance of evolutionary theory for reducing rape lies in its ability to identify likely proximate causes, which may enable individuals to eliminate the immediate factors that bring about rape. Therefore, we have emphasized the importance of identifying the developmental cues that construct the adaptation responsible for rape as well as the cues that activate this adaptation after its ontogenetic construction.

Why is rape a horrendous experience for the victim?

Mate choice was a fundamental means of reproductive success for females in human evolutionary history. Thus, rapists' circumvention of mate choice has had extremely negative consequences for female reproductive success throughout human evolutionary history. The psychological pain that rape victims experience appears to be an evolved defense against rape. The pain focuses the victim on the rape and on the negative changes it has brought about in her life, thus helping her to solve current problems (e.g., her mate's divestment and suspicions) and to avoid being raped again. Women also appear to have psychological adaptations other than mental pain that also defend against rape, such as avoiding contexts with elevated risk of rape when at the point of maximum fertility in the menstrual cycle.

The females who outreproduced others and thus became our ancestors were individuals who were highly distressed by rape.

Why does the mental trauma of rape vary with the victim's age and marital status?

From biological (evolutionary) theory, we can predict that the negative reproductive consequences faced by rape victims in evolutionary history depended on their age and on their pair-bond status. Since only reproductive-age females can get pregnant from rape, females of reproductive age and those with investing mates are predicted to have experienced the greatest negative consequences. Males of species in which males engage in parental effort have been selected to invest in their own offspring because cuckoldry was, in human evolutionary history, a persistent problem that lowered or eliminated a male's offspring production. A female with an investing mate faced the prospect of losing some or all of the mate's investment as a result of his concern about rape's effect on his paternity. Research shows that females of reproductive age and married women indeed exhibit more psychological pain after rape than other females. That mental pain is a variable response rather than an invariant one is predicted on the ground that mental pain historically entailed costs to reproductive success (such as distraction from other important life events) as well as reproductive benefits.

Why does the mental trauma of rape vary with the sex acts?

The epitome of circumvention of female mate choice is insemination by an unwanted mate. If such insemination results in pregnancy, it leads the female to expend her limited parental investment in a maladaptive manner. Research indicates that rapes that include copulation give rise to more psychological pain in female victims than non-copulatory sexual assaults. This pattern is driven by the greater pain of reproductive-age females, who during human evolutionary history had the most to lose (in terms of lowered reproductive success) from insemination by rapists.

Why does the mental trauma of rape decrease as physical injuries increase?

The answer to this question is related to men's evolved desire to invest in offspring they have sired rather than in offspring born to their mates but sired by other men. A rape is less threatening to a man's paternity than a consensual affair. That a partner has indeed been raped is less ambiguous if there are visible signs that she resisted. Thus, evolutionary theory predicts that raped women with visible signs of having resisted will experience less post-rape mental anguish, and indeed this is a strong pattern in women's response to rape trauma.[1]

Why do young males rape more often than older males?

Sexual selection on males for mate number has endowed boys and men with a much greater risk proneness than is seen in human females. The risk proneness of male humans peaks in young adulthood (Wilson and Daly 1985). So does that of females, but they are much less risk prone than males (Campbell 1995). Both peaks stem from sexual selection's favoring maximal risk taking when the competition for entering the breeding population is most intense: at the onset of adulthood. Males have been strongly sexually selected to pursue resources and climb the social ladder at this stage because success in these pursuits positively affected male reproductive success in human evolutionary history (Alexander 1979; Weisfeld 1994). That men's sexual interest and impulsiveness peak early in adulthood is also due to past sexual selection. The combination of the two peaks—that of risk taking and that of sexual desire—accounts for the fact that young men rape more.

Why are young women more often the victims of rape than older women or girls?

Fertility is strongly related to age in women but not in men. In Western countries, menarche usually occurs between the ages of 10 and 13 (Barber 1998). Women's fertility declines markedly after age 30 and drops to zero at menopause. Men's fertility begins at puberty and remains high into the fifties or even later, with no abrupt cessation comparable to menopause. The sex difference in fertility schedules gives women a narrow window of opportunity for offspring production. This had tremendous evolutionary consequences in humans, and it led to selection on males for preferring young women as mates. This preference is manifested in men's pursuit of consensual and non-consensual sex. Young women are the focus of men's sexual interest, whether the context is prostitution, pornography, marriage, romantic affairs, or rape. Men's focus on visual indicators of youth is due to the fact that a female's mate value is largely manifested in her bodily signals of fertility. A male's mate value, in contrast, is spread across bodily features, resources, and status.

Why is rape more frequent in some situations, such as war, than in others?

Humans (including rapists), like all animals, pay attention to costs and benefits when making decisions. Rape by conquering soldiers is common because the benefits are high (many young women are available) and because the costs are low (the women are vulnerable; the rapists are anonymous and relatively free from sanctions against rape). Cost-benefit considerations also go part of the way toward explaining why modern societies have relatively high rape rates. In such societies, young women rarely are chaperoned and often encounter social circumstances that make them vulnerable to rape. Moreover, anonymity is more prevalent, and the sanctions against rape are less effective deterrents, in modern societies than in traditional human environments, where individuals are more likely to be known to others.

Why does rape occur in all known cultures?

The capacity to rape is attributable ultimately to selection and proximately to ontogeny and adaptation. Rape behavior arises from elements of men's sexual nature—their sexual psychology. This psychology is characteristic of men in general, but not of pre-pubescent boys. It is reliably generated during development across a wide range of rearing environments.

Why are some instances of rape punished in all known cultures?

Rape within the rapist's own social group (known as *in-group* rape and usually defined by kinship in traditional societies) is punished across human societies because it has negative consequences for the reproductive success of all whose reproductive interests overlap with those of the victim, including her relatives, her mate, and her mate's relatives. Sanctions against in-group rape take the form of codified rules (law) and unwritten rules that protect the interests of those who make and enforce them. The rules and other forms of teaching are directed at controlling men's sexual desires.

Why are people (especially husbands) often suspicious of the victim's claim to have been raped?

The husband of the victim is looking for evidence of an unambiguous rape that the woman tried to avoid to the best of her ability—an act that is less threatening to his paternity than either a consensual affair or a rape that the woman did not resist strongly. In his suspicions, the husband is unconsciously assessing the cost of the rape to him and whether he should desert the mate or continue to invest in her and her offspring. Significant social allies of the victim other than her husband are also gaining information about the rape's impact on their reproductive interests through their suspicions. On theoretical grounds, they are expected to adjust their investment in the victim according to the same criteria that her husband uses.

Intuition that some women use false accusations, gossip, rumor, and ostracism of others to obtain resources and other social benefits may also play a part in suspicion toward claims of rape. Indeed, women do use these tactics more than men. Men rely more on direct forms of competition, such as intimidation and aggression. Humans are socially astute because of past selection for analysis of social behavior, which contains many predictive elements that are also products of past selection.

Why is rape often treated as a crime against the victim's husband?

Men are selected to control the sexual behavior of their mates and of their sisters and daughters. Control of mates serves the purpose of protecting paternity; control of female relatives is intended to make them more attractive to discriminating men with resources. Because in human evolutionary history only males could be cuckolded, a female's mate value depended greatly on how much paternity reliability she could provide. Rape of a woman, therefore, is viewed as a cost to the men whose reproductive interests she is expected to serve.

Why have attempts to reform rape laws met with only limited success?

To date, attempts to reform rape laws have not taken into account various evolved human intuitions about human behavior. This is true of all aspects of the reform movement, including expanding the definition of rape, eliminating the requirements that rape be corroborated by other persons and that proof of the victim's resistance be provided, and disallowing consideration of the victim's sexual history.

Humans are selected to define rape in a specific way because, as discussed just above, the events involved in the sexual act affect the woman's value to her male and female social allies. Much the same is true of rape corroboration, victim resistance, and victim sexual history. Furthermore, intuition about women's use of sexual allegations for self-gain may tend to block reform of the laws.

Why does rape exist in many, but not all, other species?

The vast majority of species show the same sex differences in sexuality that humans show: relative to females, males are more eager to mate, show less discrimination about mating partners, and pursue sexual variety without commitment. Females, on the other hand, are very choosy about mates. All these species have an evolutionary history of a high degree of polygyny. A very small percentage of species practice monogamy, but even in these species the stronger sexual selection for high mate number (attributable to the trivial minimum investment by males) creates virtually the same male sexual features seen in highly polygynous species, with "monogamous" males seeking extra-pair copulation. (A fraction of species show sex-role reversal, the males choosing mates and the females competing for multiple mates. Sex-role reversal arises out of the rare circumstance in which males contribute more parental investment than do females.)

Although males of essentially all species exhibit sexual psychological adaptations for obtaining a high number of mates, rape is not universal across animal species. It is, however, common. Most of the attempts to theorize about the ecological variation that creates selection for rape are very recent. Recent research by biologists examining selection for and against rape focuses on ecological factors that affect rape's benefits and costs, such as whether or not females are distributed spatially in ways that make them vulnerable to males (for example, whether they form female-female social alliances against coercive males, and whether they live within range of protective relatives). Also, how well adapted females are against rape in the coevolutionary race with rapist males appears to be relevant to the selection on males to rape. Another factor is the cost of rape to females and to rapist males when the degree of force used lowers female survival (Clutton-Brock and Parker 1995).

Some of the same factors that select for or against rape across species may be relevant to the conditional use of rape by men across human societies (Smuts and Smuts 1993). Protection of women by family members and husbands may be especially important in inhibiting rape; structural barriers such as chaperoning may also help.

Why does rape still occur among humans?

In chapter 1 we suggested two reasons for why it has not yet proved possible to eliminate rape: First, social scientists and people in general have such a limited understanding of ultimate causation that they fail to see how evolutionary theory can contribute to an understanding of rape's proximate causes. Dawkins (1986), Alexander (1987) and Pinker (1997) have discussed some possible reasons for this difficulty, including ideological biases and the fact that thinking about ultimate causation requires us to envision vast stretches of historical time when we are designed only to understand short-term causal factors. We have emphasized the role of ideological bias, but certainly there is reason to think that other factors often play important roles. Second, most attempts to eliminate rape have been based on the social science explanation, which is fundamentally ideological rather than empirical and which hence contains many fundamental inaccuracies.

Biologists are in a position to inform others about how evolution applies to humans. Biologists have deep understanding of ontogeny and of the theory of evolution, both of which deny the dichotomies for traits of individuals (e.g., learned vs. genetic) that have misinformed and misguided social scientists' research on rape. In addition, the existence of rape across many animal species is one of the many evidentiary blows biology has dealt to the social science explanation, which insists that rape is not natural, not evolved, not biological, uniquely human, and attributable to "culture" (a non-biological entity that works its magic through children's arbitrary socialization). The existence of the same differences between the sexes in humans and in most other animals is another blow to the notion that rape is just about cultural causation, since relatively few of the non-human species have any socialization of offspring at all, much less sex-differentiated socialization.

Recently some social scientists have asserted that the entire "animal literature" does not count. Polaschek et al. (1997, p. 128) suggest that this literature is "seriously flawed in both what is observed and the interpretations made" and that it reveals "a male-centered view." Polaschek et al. cite Gowaty's 1992 paper "Evolutionary biology and feminism" as supporting their claim. In a few sentences they discount all evidence for rape in non-human species and dismiss all the data on the evolution of sexual differences[2]—findings that, in fact, are clearly relevant to an understanding of human behavior, as we have explained. Polaschek et al. go on to criticize the evolutionary view of rape on the ground that men's strong desire for partner number is cultural rather than biological. This too reflects a fundamental misunderstanding of the fact that culture is biology. We would not be surprised to hear social scientists suggest, next, that insects, other arthropods, other invertebrates, and most vertebrates are somehow being influenced during development by music videos, television, and movies.

How can rape be prevented?

Among the important implications of the biological understanding of rape is that programs meant to educate people about rape should be revised so as to stop spreading the notion that rape is not sex but violence. Both young men and young women should be taught about male and female sexuality (particularly male sexuality) and about risk factors for rape. Men should be informed of the penalties for rape. Structural barriers to rape should be considered, and efforts to change evolved biases about rape accusations should be considered. Finally, changes in the law with regard to rape and related matters should be based on scientific knowledge.

Though we have not advocated specific methods of punishing rapists, we have stressed the value of punishment for changing human behavior, keeping in mind that some claims of rape are false and that false conviction is possible. Voters must decide what is a suitable punishment for rape. Science has nothing to say about what is right or wrong in the ethical sense. Biology provides understanding, not justification, of human behavior. Biological knowledge is useful to a democratic society to the extent that it can be used to achieve goals that people decide are appropriate. These goals are typically based on ideological considerations. Whereas many ideological issues (e.g., abortion, conservation, taxation) involve a great deal of disagreement, it is safe to say that the vast majority of people are against rape. This being the case, it is our hope that concerned people will begin making use of the knowledge that evolutionary biology provides in order to reduce the incidence of rape and to better deal with this horrendous crime's effects on its victims and their significant others.

Notes

Preface

1. See, e.g., Ruse 1994; Holton 1993, 1996; Gross and Levin 1994; Wilson 1998. (Throughout the book, immediately relevant references are cited in text; supplementary references, such as these, are cited in notes, which the general reader may skip.)

2. For further discussion, see Alexander 1988. On rhetoric and the humanities, see Raymond 1982.

Chapter 1

1. See also Estrich 1987, pp. 58–59.

2. For purposes of the survey, this was defined as penetration by a penis, fingers, or an object.

3. See also Williams 1988.

4. For detailed critiques, see Alexander and Borgia 1978; Palmer et al. 1997; Maynard Smith 1998; Nunney 1998; Trivers 1999.

5. See glossary for definitions.

6. Of course, the functional design of fox feet remains an empirical question. In the future, anatomists may find features of fox feet that are designed to pack snow.

7. See also Curio 1973; Dewsbury 1980; Palmer 1988b, 1989b; Symons 1987a,b; Crawford 1993.

8. Sexual selection will be defined and elaborated on in chapter 2.

9. See also Sulloway 1996.

10. See also Ellis 1992; Thornhill and Gangestad 1993, 1996; Buss 1994; Jones 1996; Thornhill and Grammer 1999.

11. See, e.g., Shepard 1992.

12. See also Symons 1979.

13. We refer readers who will believe such evidence only when they see it to page 51 of David Marr's 1982 book *Vision*. Alcock (1993, p. 153) describes the optical illusion illustrated there as follows: "(A) We see illusory gray spots at the intersections of the white bars. Our brains exaggerate the contrast wherever there is an edge (black bordered by white), and therefore white areas without edges appear darker than they are. (B) We perceive white forms overlying the triangles—but the forms that we see do not exist." This optical illusion, Alcock observes (p. 153), "tells us that our visual mechanisms, like those of toads, have special features that 'encourage' us to detect specific stimuli. Certain of our visual cells react exceptionally strongly to edges, actually enhancing the contrast between neighboring dark and light areas. There is no more light reaching our eyes from the white paper right next to the black squares than from the central grayish regions that are not bordered directly by black. But our brain tells us that there is, creating a useful illusion of the sort that contributes to our ability [and contributed to the ability of our ancestors living in past environments] to see the outlines of objects."

14. Cognitive adaptations function to make inferences about the circumstances around or inside an animal. Cognition pertains to what the animal knows—its "intelligence." In humans, at least some of the inferences have conscious components (Pinker 1997).

15. See Flinn and Alexander 1982; Daly 1982; Flinn 1997.

16. See also Daly 1982; Flinn and Alexander 1982; Tooby and Cosmides 1989; Dennett 1995.

Chapter 2

1. See, e.g., Low 1978; Alexander 1979.

2. See also Bateman 1948 and pp. 183–186 of Williams 1966.

3. See Thornhill and Alcock 1983; Trivers 1985; Thornhill 1986; Thornhill and Gwynne 1986; Alcock 1997. Note that Trivers's theory addresses the extent rather than the nature of sexual selection. The evolution in animal species of the many ways by which the members of the limited sex compete (e.g., by searching widely for mates, by defending resources critical to their mates, by defending mates directly, or by escalated intrasexual aggression) is a separate issue (for appropriate theory and further discussion, see Trivers 1972; Bradbury and Vehrencamp 1977; Emlen and Oring 1977; Thornhill and Alcock 1983).

4. The citations that follow these predictions refer to the individual(s) who derived the respective predictions from evolutionary theory.

5. For predicted exceptions, see Alexander et al. 1979.

6. For predictable exceptions involving mammals in which male aggression is counter-reproductive and thus has not evolved, see Schwagmeyer 1988.

7. The following references are the most relevant for supplying comparative data (including that for humans) demonstrating the truth of the predictions:

Darwin 1874; Alexander 1979, 1987; Alexander et al. 1979; Symons 1979; Clutton-Brock et al. 1982; Daly and Wilson 1983, 1988; Trivers 1985; Rubenstein and Wrangham 1986; Weisfeld and Billings 1988; Grammer 1993; Buss 1994; Weisfeld 1994; Geary 1998; Townsend 1998; Ghiglieri 1999.

8. See, e.g., Alexander et al. 1979; Wittenberger 1981; Daly and Wilson 1983; Thornhill and Alcock 1983; Trivers 1985.

9. In this context, 'reported' means *reported to police.*

10. For further discussion and for evidence of human sperm competition, see Baker and Bellis 1995.

11. See also Townsend and Levy 1990; Ellis 1992; Grammer 1993; Perusse 1993; Townsend 1998. Buss's studies on human mate preference, and studies by others, are summarized in Buss's popular book *The Evolution of Desire* (1994).

12. For a review, see Andersson 1994.

13. For reviews, see Andersson 1994; Møller and Thornhill 1998a,b; Møller and Alatalo 1999).

14. For a review, see Møller and Swaddle 1997.

15. A moderate amount of variation among individuals in symmetry is due to individual genetic differences; the rest is environmental (Møller and Thornhill 1997; Gangestad and Thornhill 1999).

16. Studies of the role of symmetry in men's sexual attractiveness to women have been done among Americans, among Europeans, and among rural Mayans in Belize. See, e.g., Gangestad and Thornhill 1997a,b; R. Baker 1997; Waynforth 1998.

17. In these studies, "romantic relationship" refers to a dating relationship of at least 3 months or a marital relationship.

18. It appears, however, that women's symmetry may affect the likelihood of marriage. In women, breast symmetry is correlated positively with marriage (Manning et al. 1997). It is reasonable to predict that women's symmetry will correlate also with the resource holdings of their mate in long-term relationships, since a woman's attractiveness is known to be a major predictor of whether she will marry a high-status man.

19. Coevolution: the simultaneous evolution in two interacting parties (species or sexes) in which each party is the cause of the evolutionary change in the other. Antagonistic coevolution occurs when one party's adaptations are maladaptive for the other party. In host-infectious disease, coevolutionary races hosts evolve defenses and disease organisms evolve adaptations to penetrate the defenses, and this goes on more or less continuously through time.

20. It may be that their androgen levels differ, and that the androgen level of each man is incorporated into his scent. In this scenario, symmetric men have higher androgen levels (perhaps because only they can afford the physiological costs).

21. For reviews, see Singh 1993; Thornhill and Gangestad 1993; Symons 1995; Barber 1995; Thornhill and Møller 1997. Perrett et al. (1998) report that women in their study found men's faces that were slightly feminized more attractive than

men's faces that were highly masculine. Highly masculine faces show greater effects of testosterone. This is interpreted by the researchers as a female preference for men who will invest in women. However, the same research group found that women who are not on the pill (i.e., are having ovulatory cycles) and are at the fertile point of their cycle prefer the most masculinized faces (Penton-Voak et al. 1999). Thus, women in whom conception is likely (i.e., women who are ovulating and at mid-cycle) prefer visual markers of male genetic quality. The same pattern is seen in women's scent preference for male symmetry.

22. For a review, see Penn and Potts 1998.

Chapter 3

1. All three forms of sexual coercion occur across many animal species (Smuts 1992; Smuts and Smuts 1993; Clutton-Brock and Parker 1995).

2. Dobzhansky et al. 1977, frontispiece.

3. See chapter 1 above. For further discussion see Williams 1966, 1992; Bell 1997.

4. Biologists believe that schizophrenia in just 1 percent of the human population requires that the genes involved be favorably selected at least in some environments at some times (Nesse and Williams 1994).

5. For cross-cultural reviews, see Palmer 1989a; Rozée 1993; Smuts and Smuts 1993.

6. See also MacKinnon 1979.

7. A form of the dominance explanation was first proposed by Susan Brownmiller. Gowaty and Buschhaus reference her idea and celebrate it on p. 219 of their 1998 paper. We discuss the ideological attraction of Brownmiller's explanation of human rape in chapter 6.

8. However, Baker and Bellis (1995) provide some evidence that men's masturbation may be an adaptation to ensure a fresh supply of sperm.

9. See also Malamuth 1996.

10. See Thornhill 1980, 1984.

11. The relationship between social disfranchisement and crime, especially in males, is strong support for stronger sexual selection on males than females in human evolutionary history. See Alexander 1979.

12. See also Malamuth and Heilmann 1998; Figueredo et al. 1999.

13. See also data for 1995 and 1996 made available by the Arkansas Crime Information Center (http://www.acic.org/1996rape.html).

14. For a review of rape of young women in various war settings, see Niarcos 1995; also see Brownmiller 1975. For the Vietnam War, see Baker 1981 and Shields and Shields 1983; for Bosnia, see Niarcos 1995 and Thompkins 1995; for Rwanda, see McKinley 1996; for World War II, see Cheng 1997.

15. Felson and Krohn (1990) discuss some of the deficiencies of the National Crime Victimization Survey data collection in this respect—deficiencies that the surveyors have attempted to deal with in recent studies.

16. See also Felson and Krohn 1990; Figueredo and McCloskey 1993; Jacobson and Gottman 1998.

17. For additional discussion of human heritable variation, see Tooby and Cosmides 1990b and Gangestad 1997.

18. See also Lalumiére and Seto 1998.

Chapter 4

1. For reviews of the literature, see Thornhill and Thornhill 1983, 1991, 1992a; Ellis 1989; Resick 1993; Gill 1996; Darvesbornoz 1997; Wilson et al. 1997.

2. See the preceding note.

3. For further details of the sample, see McCahill et al. 1979; see also Thornhill and Thornhill 1990a,b,c, 1991.

4. Girls younger than 11 and women over 45 have very low fertility in the United States (Thornhill and Thornhill 1983). Fertility refers to age-specific birth rate and thus is related to the probability that any given copulation will lead to pregnancy, gestation, and live birth.

5. Katz and Mazur (1979) report the same. See also Bownes et al. 1991.

6. For further discussion and a partial review of this literature, see Nesse and Williams 1994.

7. For reviews of recent research, see Resick 1993 and Thornhill 1997b.

8. Raped pregnant horses sometimes abort.

9. For a discussion of R. Thornhill's results on ejaculate dumping by jungle fowl hens, see Birkhead and Møller 1992.

10. See also Whitaker 1987.

11. See also Baker and Bellis 1995.

12. For discussions of such female-female alliances, see Smuts and Smuts 1993 and Mesnick 1997.

Chapter 5

1. For discussions of salient work done during this period, see Hamilton 1963, 1964; Williams 1966, 1985, 1992; Alexander 1971, 1974, 1975, 1987; Trivers 1971, 1972, 1974, 1985; Alcock 1975, 1997; Brown 1975; Wilson 1975; Dawkins 1976, 1986; Zahavi and Zahavi 1997; Maynard Smith 1998.

2. See also Snowden 1997.

3. See also Sorenson and White 1992.

4. See also Gould 1997; Avise 1998.

5. There is no example of true altruism adaptation known in biology. This is the kind of adaptation that evolution by group selection for true altruism would produce if it had been effective in life's evolutionary history. A biologist who discovered a true altruism adaptation would become a famous biologist because of the novelty of such a finding. Biologists in the mid 1960s and the early 1970s flocked to study cases of possible true altruism discussed by Williams (1966). They all failed to find evidence for true altruism adaptation; instead, they found evidence of design by individual selection, and the stamp of effective individual selection is seen in the multitude of adaptations of each of the several million species known to biologists.

6. See Allen et al. 1975; Sociobiology Study Group 1978.

7. See, e.g., Mayr 1983; Thornhill and Alcock 1983; Dawkins 1986; Alexander 1987; Wright 1990, 1994; Dennett 1995; Maynard Smith 1995; Queller 1995; Pinker 1997; Alcock 1998.

8. See also Williams 1966, 1992; Buss et al. 1998.

9. See Queller 1995.

10. See also Mesnick 1997.

11. See also Zillmann 1998.

12. See also Dusek 1984; Blackman 1985; Schwendinger and Schwendinger 1985; Sunday 1985.

13. For a detailed discussion of the kinds of comparisons that are valid in science, especially in regard to so-called extrapolation, see Thornhill et al. 1986.

Chapter 6

1. For further discussion and a detailed critique of learning theory, see Tooby and Cosmides 1992.

2. On the history of this movement, see Freeman 1983 and Brown 1991.

3. For other discussions of this combination of learning theory and feminist ideology, see Palmer 1988a; Ellis 1989; also Symons 1979; Russell 1982, 1984; Shields and Shields 1983; Thornhill and Thornhill 1983; Stock 1991.

4. See also Drake 1990; Herman 1990.

5. For other examples, see the articles in Swisher and Wekesser 1994.

6. See also MacKinnon 1987, p. 6; Dworkin 1989, p. 165; Southern Women's Writing Collective 1990, p. 143.

7. For more on human violence in evolutionary contexts, see Daly and Wilson 1988; Palmer 1993; Furlow et al. 1998. Also see our discussion, in chapters 2 and 3, of men's violence directed at mates.

8. Data on the ages of rapists are voluminous. The recent data are contained in the same crime surveys we discussed in chapter 3 in connection with rape victims' ages. Older data sets are discussed in Thornhill and Thornhill 1983.

9. For reviews of military rape, see Littlewood 1997 and Morris 1996. Hartung (1992) cites the vivid description found in Numbers 31:1–35.

10. See also Shields and Shields 1983, p. 121.

11. See also Russell 1975; Geis 1977; Queen's Bench Foundation 1978; Rada 1978a; Katz and Mazur 1979; Sussman and Bordwell 1981; Ageton 1983.

12. See also Groth 1979, pp. 50, 55, 93, 159, 161, 181, 183.

13. See also Gebhard et al. 1965; Amir 1971; Schiff 1971; Burgess and Holmstrom 1974; Chappell and Singer 1977; Felson and Krohn 1990.

14. See also Gebhard et al. 1965.

15. The minimization of violence-related costs by the rapists, however, doesn't itself identify rape-specific adaptation; it may be attributable to adaptation for using violence to control the behavior of other people in general.

16. See also Wrangham and Peterson 1996.

17. See also Smuts 1992; Smuts and Smuts 1993.

18. See also Symons 1979; Palmer et al. 1997.

19. See also Palmer 1989b; Wrangham and Peterson 1996.

Chapter 7

1. Students have often asked us this question.

2. According to John Hartung, an expert on the Bible and related documents, the Codes of Maimonides are the foundation of all Western legal systems (personal communication).

3. See also Sterling 1995; Furnham and Brown 1996; Willis and Wrightsman 1995.

4. Beckstrom (1993) discusses punishment for rape in some detail in his pioneering book *Darwinism Applied.*

5. For a generally useful discussion of how sanctions might be used to punish and reduce rape during war, see Green et al. 1994.

Chapter 8

1. For example, Fonow et al. (1992, p. 115) report finding in their educational program that "the men were . . . more likely than the women to agree with the statement that rape is for sex," and that "thus, the men in this study were more likely to believe the myths grounded in the idea of rape as sex." See also Ward 1995.

2. For more on the effects of evolved developmental cues on men's sexual behavior, see Thornhill and Thornhill 1983, 1992a,b.

Chapter 10

1. For a detailed discussion of defensive tactics that are useful against rapists, see Booher 1991.

Chapter 11

1. See also see Rada 1978a,c.

Chapter 12

1. This pattern was first observed by McCahill et al. (1979), who did not know what to make of it.
2. A considerable amount of this evidence has been gathered by female biologists.

Glossary

adaptation
a bodily trait that is a product of direct selection for the adaptation's function (includes psychological traits)

biological
living or once living, and hence a developmental product of gene-environment causal interactions

biparental
adjective referring to a mateship circumstance in some species (e.g., humans and many birds) in which a male parent and a female parent form a pair bond and both invest in the offspring

coercion
any form of force or influence involving the application or the threat of a negative consequence (see also *sexual coercion*)

consciousness
the state of being aware of or feeling the physiological events of mentation (mental activity)

cuckold
an individual who unknowingly invests in an offspring that is genetically related to his mate but not to him (almost solely applied to males in biparental species)

drift
chance variation in reproductive success among individuals

facultative
dependent on specific environmental variables; equivalent to *condition dependent*

gene flow
movement of genes from one population to one or more others by movement of individuals between populations

heritability
the degree to which differences between individuals are attributable to differences in genes

maladaptive trait
a trait whose costs exceed its benefits, cost and benefits being measured in terms of reproductive success

mate
(verb) to have sexual intercourse with; (noun) an individual with which another has sexual intercourse

mutation
a change in a gene that can be transmitted between generations

ontogeny
the processes involved in the development of the traits of an individual within its lifetime

pair bond
an extended attachment between two individuals who are mates or potential mates

paternity reliability
a male's confidence that he is the genetic sire of a mate's offspring

phenotypic
adjective referring to bodily features

phylogenetic
adjective referring to relatedness of organisms attributable to descent from common ancestral species

polygyny
a mating system in which sexual selection acts more strongly on males than on females

psyche
the brain

rape
copulation resisted to the best of the victim's ability unless such resistance would probably result in death or serious injury to the victim or in death or injury to individuals the victim commonly protects

reproductive value
the ability of an individual organism to contribute offspring to the population in the future

romantic relationship
a pair-bond relationship involving mutual sexual interest and feelings of non-nepotistic love

sexual coercion
coercion whose goal is to gain sexual stimulation from another

sexual selection
the selection of traits that increase the quantity and/or the quality of an individual's mates

social construction
the view that the reified abstraction "society" constructs and accounts for human behavior and psychological features

social science explanation
applied to rape, the view that it is essentially or completely the result of capricious learning

social science model
the hypothesis that human behavior and psychology are only or primarily the result of learning experience that is arbitrary in nature

unreported
not reported to police

References

Afton, A. 1985. Forced copulation as a reproductive strategy of male lesser Scaup *Aythya affinis*: A field test of some predictions. *Behaviour* 92: 146–167.

Ageton, S. 1983. *Sexual Assault among Adolescents*. Lexington.

Ahmad, Y., and P. Smith. 1994 Bullying in schools and the issue of sex differences. In *Male Violence*, ed. J. Archer. Routledge.

Alcock, J. 1975. *Animal Behavior*. Sinauer.

Alcock, J. 1993. *Animal Behavior: An Evolutionary Perspective*, fifth edition. Sinauer.

Alcock, J. 1997. *Animal Behavior: An Evolutionary Perspective*, sixth edition. Sinauer.

Alcock, J. 1998. Unpunctuated equilibrium in the natural history essays of Gould, Stephen Jay. *Evolution and Human Behavior* 19: 321–336.

Alexander, R. 1971. The search for an evolutionary philosophy of man. *Proceedings of the Royal Society of Victoria* 84: 99–120.

Alexander, R. 1974. The evolution of social behavior. *Annual Review of Ecology and Systematics* 5: 325–383.

Alexander, R. 1975. The search for a general theory of behavior. *Behavioral Sciences* 20: 77–100.

Alexander, R. 1978. Evolution, creation and biology teaching. *American Biology Teacher* 40: 91–107.

Alexander, R. 1979. *Darwinism and Human Affairs*. University of Washington Press.

Alexander, R. 1987. *The Biology of Moral Systems*. Aldine de Gruyter.

Alexander, R. 1988. The evolutionary approach to human behavior: What does the future hold? In *Human Reproductive Behavior*, ed. L. Betzig et al. Cambridge University Press.

Alexander, R. 1989. The evolution of the human psyche. In *The Human Revolution*, ed. P. Mellars and C. Stringer. University of Edinburgh Press.

Alexander, R. 1990. Epigenetic rules and Darwinian algorithms: The adaptive study of learning and development. *Ethology and Sociobiology* 11: 241–303.

Alexander, R., and G. Borgia. 1978. Group selection, altruism, and the levels of organization of life. *Annual Review of Ecology and Systematics* 9: 449–474.

Alexander, R., and K. Noonan. 1979. Concealment of ovulation, parental care, and human social evolution. In *Evolutionary Biology and Human Social Behavior*, ed. N. Chagnon and W. Irons. Duxbury.

Alexander, R., J. Hoogland, R. Howard, K. Noonan, and P. Sherman. 1979. Sexual dimorphisms and breeding systems in pinnipeds, ungulates, primates, and humans. In *Evolutionary Biology and Human Social Behavior*, ed. N. Chagnon and W. Irons. Duxbury.

Allen, G., and L. Simmons. 1996. Coercive mating, fluctuating asymmetry and male mating success in the dung fly, *Sepsis cynipsea*. *Animal Behaviour* 52: 737–741.

Allen et al. 1975. Against "sociobiology." *New York Review of Books*, November 13: 182–185.

Allgeier, E., and A. Allgeier. 1991. *Sexual Interactions*, third edition. Heath.

Amir, M. 1971. *Patterns in Forcible Rape*. University of Chicago Press.

Andersen, N. 1997. A phylogenetic analysis of the evolution of sexual dimorphism and mating systems in waterstriders (Hemiptera, Gerridae). *Biological Journal of the Linnaean Society* 61: 345–368.

Andersson, M. 1994. *Sexual Selection*. Princeton University Press.

Arnqvist, G. 1989. Sexual selection in a water strider: The function, mechanism and selection and heritability of a male grasping apparatus. *Oikos* 56: 344–350.

Arnqvist, G. 1992. Spatial variation in selective regimes: Sexual selection in the water strider, *Gerris odontogaster*. *Evolution* 46: 914–929.

Arnqvist, G., and L. Rowe. 1995. Sexual conflicts and arms races between the sexes: A morphological adaptation for control of mating in a female insect. *Proceedings of the Royal Society of London* B 261: 123–127.

Avise, J. 1998. *The Genetic Gods: Evolution and Belief in Human Affairs*. Harvard University Press.

Bachman, R., and L. Saltzman. 1995. Violence against Women: Estimates from the Redesigned Survey. Special Report, Bureau of Justice Statistics, US Department of Justice.

Bailey, J., S. Gaulin, Y. Agyei, and B. Gladue. 1994. Effects of gender and sexual orientation on evolutionarily relevant aspects of human mating psychology. *Journal of Personality and Social Psychology* 66: 1081–1093.

Bailey, R., R. Seymour, and G. Stewart. 1978. Rape behavior in blue-winged teal. *Auk* 95: 188–190.

Baker, K. 1997. Once a rapist? Motivational evidence and relevance in rape law. *Harvard Law Review* 110: 563–624.

Baker, M. 1981. *Nam: The Vietnam War in the Words of the Soldiers Who Fought There*. Berkeley.

Baker, R. 1997. Copulation, masturbation and infidelity: State-of-the-art. In *New Aspects of Human Ethology*, ed. A. Schmitt et al. Plenum.

Baker, R., and M. Bellis. 1989. Number of sperm in human ejaculates varies in accordance with sperm competition theory. *Animal Behaviour* 37: 867–869.

Baker, R., and M. Bellis. 1993. Human sperm competition: Adjustment by males and the function of masturbation. *Animal Behaviour* 46: 861.

Baker, R., and M. Bellis. 1995. *Human Sperm Competition: Copulation, Masturbation and Infidelity*. Chapman and Hall.

Barash, D. 1977. Sociobiology of rape in mallards (*Anas platyrhynchos*): Responses of the mated male. *Science* 197: 788–789.

Barash, D. 1979. *The Whisperings Within*. Harper and Row.

Barber, N. 1995. The evolutionary psychology of physical attractiveness: Sexual selection and human morphology. *Ethology and Sociobiology* 16: 395–424.

Barber, N. 1998. *Parenting: Roles, Styles and Outcomes*. Nova.

Barber, N. 1994. Secular changes in standards of bodily attractiveness in women: Tests of a reproductive model. *International Journal of Eating Disorders* 23: 449–453.

Barkow, J., L. Cosmides, and J. Tooby, eds. 1992. *The Adapted Mind: Evolutionary Psychology and the Generation of Culture*. Oxford University Press.

Barlow, G. 1967. Social behavior of a South American leaf fish *Polycentrus schonburk*: With an account of recurring pseudofemale behavior. *American Middle Naturalist* 78: 215–234.

Baron, L. 1985. Does rape contribute to reproductive success? Evaluations of sociobiological views of rape. *International Journal of Women's Studies* 8: 266–277.

Bateman, A. 1948. Intrasexual selection in *Drosophila*. *Heredity* 2: 349–368.

Beckstrom, J. 1993. *Darwinism Applied: Evolutionary Paths to Social Goals*. Praeger.

Beecher, M., and I. Beecher. 1979. Sociobiology of bank swallows: Reproductive strategy of the male. *Science* 205: 1282–1285.

Bell, G. 1997. *Selection: The Mechanism of Evolution*. Chapman and Hall.

Bell, V. 1991. Beyond the "Thorny Question": Feminism, Foucault and the desexualisation of rape. *International Journal of the Sociology of Law* 19: 83–100.

Beneke, T. 1982. *Men on Rape*. St. Martin's.

Benshoof, L., and R. Thornhill. 1979. The evolution of monogamy and concealed ovulation in humans. *Journal of Social and Biological Structures* 2: 95–106.

Berger, R., P. Searles, and W. Neuman. 1988. The dimensions of rape reform legislation. *Law and Society Review* 22: 329–357.

Betzig, L. 1986. *Despotism and Differential Reproduction: A Darwinian View of History*. Aldine de Gruyter.

Betzig, L. 1989. Causes of conjugal dissolution: A cross-cultural study. *Current Anthropology* 30: 654–676.

Betzig, L. 1995. Wanting women isn't new; getting them is—very. *Journal of Politics and the Life Sciences* 14: 24.

Betzig, L., ed. 1997. *Human Nature: A Critical Reader*. Oxford University Press.

Betzig, L., M. Borgerhoff Mulder, and P. Turke, eds. 1988. *Human Reproductive Behavior: A Darwinian Perspective*. Cambridge University Press.

Bingman, V. 1980. Novel rape avoidance in mallards *Anas platyrhyncos*. *Wilson Bulletin* 92: 405.

Birkhead, T. 1979. Mate guarding in the magpie *Pica pica*. *Animal Behaviour* 33: 608–619.

Birkhead, T., and A. Møller. 1992. *Sperm Competition in Birds: Evolutionary Causes and Consequences*. Academic Press.

Birkhead, T., and A. Møller, eds. 1998. *Sperm Competition and Sexual Selection*. Academic Press.

Birkhead, T., S. Johnson, and D. Nettleship. 1985. Extra-pair matings and mate guarding in the common Murre *Uria aalge*. *Animal Behaviour* 33: 608–619.

Bjorkqvist, K., K. Osterman, and K. Langerspetz. 1994. Sex differences in overt aggression among adults. *Aggressive Behavior* 20: 27–34.

Blackman, J. 1985. The language of sexual violence: More than a matter of semantics. In *Violence against Women*, ed. S. Sunday and E. Tobach. Gordian Press.

Blatt, D. 1992. Recognizing rape as a method of torture. *Review of Law and Social Change* 19: 821–865.

Bodmer, W., and L. Cavalli-Sforza. 1976. *Genetics, Evolution, and Man*. Freeman.

Booher, D. 1991. *Rape: What Would You Do If . . . ?* Messner.

Borgia, G. 1979. Sexual selection and the evolution of mating systems. In *Sexual Selection and Reproductive Competition in Insects*, ed. M. Blum and N. Blum. Academic Press.

Bossema, I., and E. Roemers. 1985. Mating strategy including mate choice in mallards *Anas platyrhynchos*. *Ardea* 72: 147–157.

Bourque, L. 1989. *Defining Rape*. Duke University Press.

Bownes, I., E. O'Gorman, and A. Sayers. 1991. Rape: A comparison of stranger and acquaintance assaults. *Medicine, Science and the Law* 31: 102–109.

Bowyer, L., and M. Dalton. 1997. Female victims of rape and their genital injuries. *British Journal of Obstetrics and Gynaecology* 104: 617–620.

Boyd, R., and P. Richerson. 1978. A simple dual inheritance model of the conflict between social and biological evolution. *Zygon* 11: 254–262.

Boyd, R., and P. Richerson. 1985. *Culture and the Evolutionary Process*. University of Chicago Press.

Bradbury, J., and S. Vehrencamp. 1977. Social organization and foraging in emballonurid bats. III. Mating Systems. *Behavioral Ecology and Sociobiology* 2: 1–17.

Bremer, J. 1959. *Asexualization*. Macmillan.

Brodzinsky, D., S. Messer, and J. Tew. 1979. Sex differences in children's expression and control of fantasy and overt aggression. *Child Development* 50: 372–379.

Broude, G., and S. Greene. 1976. Cross-cultural codes on twenty sexual attitudes and practices. *Ethnology* 15: 409–429.

Brown, D. 1991. *Human Universals*. McGraw-Hill.

Brown, J. 1975. *The Evolution of Behavior*. Norton.

Browne, K. 1995. Sex and temperament in a modern society: A Darwinian view of the glass ceiling and the gender gap. *Arizona Law Review* 37: 971–1106.

Brownmiller, S. 1975. *Against Our Will: Men, Women, and Rape*. Simon and Schuster.

Brownmiller, S. 1976. *Against Our Will: Men, Women, and Rape*. Paperback edition. Bantam.

Brownmiller, S., and B. Mehrhof. 1992. A feminist response to rape as an adaptation in men. *Behavioral and Brain Science* 15: 381–382.

Buchwald, E., P. Fletcher, and M. Roth. 1993. Editor's preface. In *Transforming a Rape Culture*, ed. E. Buchwald et al. Milkweed.

Burgess, A., and L. Holmstrom. 1974. *Rape: Victims of Crisis*. Brady.

Burley, N. 1979. The evolution of concealed ovulation. *American Naturalist* 114: 835–858.

Burley, N., and R. Symanski. 1982. Women without: An evolutionary and cross-cultural perspective on prostitution. In *The Immoral Landscape*, ed. R. Symanski. Butterworths.

Burns, J., K. Cheng, and F. McKinney. 1980. Forced copulations in captive mallards: 1. fertilization of eggs. *Auk* 97: 875–879.

Buss, D. 1985. Human mate selection. *American Scientist* 73: 47–51.

Buss, D. 1987. Sex differences in human mate selection criteria: An evolutionary perspective. In *Sociobiology and Psychology*, ed. C. Crawford et al. Erlbaum.

Buss, D. 1989. Sex differences in human mate preferences: Evolutionary hypotheses tested in 37 cultures. *Behavioral and Brain Science* 12: 1–14.

Buss, D. 1994. *The Evolution of Desire: Strategies of Human Mating*. Basic Books.

Buss, D. 1999. *Evolutionary Psychology: The New Science of the Mind*. Allyn and Bacon.

Buss, D., and N. Malamuth. 1996. Introduction. In *Sex, Power, Conflict*, ed. D. Buss and N. Malamuth. Oxford University Press.

Buss, D., and D. Schmitt. 1993. Sexual strategies theory: An evolutionary perspective on human mating. *Psychological Reviews* 100: 204–232.

Buss, D., R. Larsen, D. Westen, and J. Semmelroth. 1992. Sex differences in jealousy: Evolution, physiology and psychology. *Psychological Science* 3: 251–255.

Buss, D., M. Haselton, T. Shackelford, A. Bleske,, and J. Wakefield. 1998. Adaptations, exaptations, and spandrels. *American Psychologist* 53: 533–548.

Byers, J. 1997. *American Pronghorn: Social Adaptations and the Ghosts of Predators Past.* University of Chicago Press.

Cade, W. 1980. Alternative male reproductive behaviors. *Florida Entomologist* 63: 30–44.

Cairns, R., B. Cairns, H. Neckerman, L. Ferguson, and J. Gariepy. 1989. Growth and aggression: 1. Childhood to early adolescence. *Developmental Psychology* 25: 320–330.

Campbell, A. 1995. A few good men: Evolutionary psychology and female adolescent aggression. *Ethology and Sociobiology* 16: 99–123.

Campbell, A., S. Muncer, and J. Odber. 1998. Primacy of organizing effects of testosterone. *Behavioral and Brain Sciences* 21: 365–380.

Caputi, J. 1993. The sexual politics of murder. In *Violence against Women*, ed. P. Bart and E. Moran. Sage.

Card, C. 1996. Rape as a weapon of war. *Hypatia* 11: 5–19.

Cashdan, E. 1993. Attracting mates: Effects of paternal investment on mate attraction strategies. *Ethology and Sociobiology* 14: 1–24.

Cashdan, E. 1996. Women's mating strategies. *Evolutionary Anthropology* 5: 134–143.

Chagnon, N., and W. Irons, eds. 1979. *Evolutionary Biology and Human Social Behavior: An Anthropological Perspective.* Duxbury.

Chappell, D., and S. Singer. 1977. Rape in New York City. In *Forcible Rape*, ed. D. Chappell et al. Columbia University Press.

Chavanne, T., and G. Gallup Jr. 1998. Variation in risk taking behavior among female college students as a function of the menstrual cycle. *Evolution and Human Behavior* 19: 1–6.

Cheng, I. 1997. *The Rape of Nanking: The Forgotten Holocaust of World War II.* Basic Books.

Cheng, K., J. Burns, and F. McKinney. 1983a. Forced copulation in captive mallards *Anas platyrhynchos* 2. Temporal factors. *Animal Behaviour* 30: 695–699.

Cheng, K., J. Burns, and F. McKinney. 1983b. Forced copulation in captive mallards *Anas platyrhynchos* 3. Sperm Competition. *Auk* 100: 302–310.

Clark, L., and D. Lewis. 1977. *Rape: The Price of Coercive Sexuality.* Women's Press.

Clutton-Brock, T., and G. Parker. 1995. Sexual coercion in animal societies. *Animal Behaviour* 49: 1345–1365.

Clutton-Brock, T., F. Guinness, and S. Albon. 1982. *Red Deer: Behavior and Ecology of Two Sexes.* University of Chicago Press.

Cohen, M., R. Garofalo, R. Boucher, and T. Seghorn. 1971. The psychology of rapists. *Seminars in Psychiatry* 3: 307–327.

Constantz, G. 1975. Behavioral ecology in the male Gila topminnow, *Poeciliopsis occidentalis* (Cyprinodontiformes: Poeciliidae). *Ecology* 56: 966–973.

Cooper, W. 1985. Female residency and courtship intensity in a territorial lizard *Holbrookia propinqua*. *Amphibia and Reptilia* 6: 63–71.

Cosmides, L., and J. Tooby. 1987. From evolution to behavior: Evolutionary psychology as the missing link. In *The Latest on the Best*, ed. J. Dupré. MIT Press.

Cosmides, L., and J. Tooby. 1989. Evolutionary psychology and the generation of culture, Part II: Case study: A computational theory of social exchange. *Ethology and Sociobiology* 10: 51–98.

Cosmides, L., and J. Tooby. 1992. Cognitive adaptations for social exchange. In *The Adapted Mind*, ed. J. Barkow et al. Oxford University Press.

Cowan, G., and R. Campbell. 1995. Rape and causal attitudes among adolescents. *Journal of Sex Research* 32: 145–153.

Cox, C., and B. Le Boeuf. 1977. Female incitation of male competition: A mechanism of sexual selection. *American Naturalist* 111: 317–335.

Crawford, C. 1993. The future of sociobiology: Counting babies or studying proximate mechanisms. *Trends in Evolution and Ecology* 8: 183–186.

Crawford, C., and B. Galdikas. 1986. Rape in nonhuman animals: An evolutionary perspective. *Canadian Psychology* 27: 215–230.

Crawford, C., and D. Krebs, eds. 1998. *Handbook of Evolutionary Psychology: Ideas, Issues and Applications*. Erlbaum.

Crespi, B. 1986. Territoriality and fighting in a colonial thrips *Hoplothrips-pedicularius* and sexual dimorphism in Thysanoptera. *Ecological Entomologist* 11: 119–130.

Crick, N., and J. Grotpeter. 1995. Relational aggression, gender and social-psychological adjustment. *Child Development* 66: 710–722.

Cronk, L. 1995. Is there a role for culture in human behavioral ecology? *Ethology and Sociobiology* 16: 181–205.

Curio, E. 1973. Towards a methodology of teleonomy. *Experientia* 29: 1045–1058.

Daly, M. 1982. Some caveats about cultural transmission models. *Human Ecology* 10: 401–408.

Daly, M., and M. Wilson. 1983. *Sex, Evolution and Behavior*, second edition. Duxbury.

Daly, M., and M. Wilson. 1988. *Homicide*. Aldine de Gruyter.

Daly, M., and M. Wilson. 1995. Discriminative parental solicitude and the relevance of evolutionary models to the analysis of motivational systems. In *The Cognitive Neurosciences*, ed. M. Gazzaniga. MIT Press.

Daly, M., and M. Wilson. 1996. Evolutionary psychology and marital conflict. In *Sex, Power, Conflict*, ed. D. Buss and N. Malamuth. Oxford University Press.

Daly, M., M. Wilson, and S. Weghorst. 1982. Male sexual jealousy. *Ethology and Sociobiology* 3: 11–27.

Darvesbornoz, J. 1997. Rape-related psychotraumatic syndromes. *European Journal of Obstetrics, Gynecology and Reproductive Biology* 71: 59–65.

Darwin, C. 1872. *The Origin of Species*. Reprint: Penguin, 1974.

Darwin, C. 1874. *The Descent of Man and Selection in Relation to Sex*. Rand McNally.

Davies, K. 1997. Voluntary exposure to pornography and men's attitudes toward feminism and rape. *Journal of Sex Research* 34: 131–137.

Dawkins, R. 1976. *The Selfish Gene*. Oxford University Press.

Dawkins, R. 1986. *The Blind Watchmaker*. Norton.

Dean, C., and M. de Bruyn-Kopps. 1982. *The Crime and Consequences of Rape*. Thomas.

Denmark, F., and S. Friedman. 1985. Social psychological aspects of rape. In *Violence against Women*, ed. S. Sunday and E. Tobach. Gordian Press.

Dennett, D. 1995. *Darwin's Dangerous Idea: Evolution and the Meanings of Life*. Simon and Schuster.

Dewsbury, D. 1980. Methods in the two sociobiologies. *Behavioral and Brain Sciences* 3 (2): 171–214.

Dickemann, M. 1979a. Female infanticide, reproductive strategies, and social stratification: A preliminary model. In *Evolutionary Biology and Human Social Behavior*, ed. N. Chagnon and W. Irons. Duxbury.

Dickemann, M. 1979b. The ecology of mating systems in hypergynous dowry societies. *Biology and Social Life* 18: 163–195.

Dickemann, M. 1981. Paternal confidence and dowry competition: A biocultural analysis of purdah. In *Natural Selection and Social Behavior*, ed. R. Alexander and D. Tinkle. Chiron.

Dietz, P. 1978. Social factors in rapist behavior. In *Clinical Aspects of the Rapist*, ed. R. Rada. Grune and Stratton.

Dobzhansky, T., F. Ayala, G. Stebbins, and J. Valentine, eds. 1977. *Evolution*. Freeman.

Donat, P., and J. D'Emilio. 1992. A feminist redefinition of rape and sexual assault: Historical foundations and change. *Journal of Social Issues* 48: 9–22.

Drake, J. 1990. Sexual aggression: Achieving power through humiliation. In *Handbook of Sexual Assault*, ed. W. Marshall et al. Plenum.

Durkheim, E. 1912. Les formes élémentaires de la vie religeuse. Reprinted in Durkheim, *The Elementary Forms of the Religious Life* (Collier, 1961).

Dusek, V. 1984. Sociobiology and rape. *Science for the People* 16: 10–16.

Dworkin, A. 1989. *Pornography: Men Possessing Women*. Dutton.

Dworkin, A. 1990. Resistance. In *The Sexual Liberals and the Attack on Feminism*, ed. D. Leidholdt and J. Raymond. Pergamon.

Eberhard, W. 1985. *Sexual Selection and Animal Genitalia*. Harvard University Press.

Eberhard, W. 1996. *Female Control: Sexual Selection by Cryptic Female Choice*. Princeton University Press.

Ehrenreich, B., and J. McIntosh. 1997. The new creationism: Biology under attack. *The Nation*, June 9.

Eisenhower, M. 1969. To Establish Justice, to Insure Domestic Tranquility. Final Report of the National Commission on Cause and Prevention of Violence. US Government Printing Office.

Ellegren, H., and A. Fridolfsson. 1997. Male-driven evolution of DNA sequence. *Nature Genetics* 17: 183–184.

Ellis, B. 1992. The evolution of sexual attraction: Evaluative mechanisms in women. In *The Adapted Mind*, ed. J. Barkow et al. Oxford University Press.

Ellis, L. 1989. *Theories of Rape: Inquires into the Causes of Sexual Aggression*. Hemisphere.

Ellis, L. 1991. The drive to possess and control as a motivation for sexual behavior: Applications to the study of rape. *Social Science Information* 30: 633–675.

Elwood, R., K. Wood, M. Gallagher, and J. Dick. 1998. Probing motivational state during agonistic encounters in animals. *Nature* 393: 66–68.

Emlen, S., and L. Oring. 1977. Ecology, sexual selection and the evolution of mating systems. *Science* 197: 215–222.

Emlen, S., and P. Wrege. 1986. Forced copulations and intra-specific parasitism: Two costs of social living in the white-fronted bee-eater. *Ethology* 71: 2–29.

Estep, E., and K. Bruce. 1981. The concept of rape in non-humans: A critique. *Animal Behaviour* 29: 1272–1273.

Estrich, S. 1987. *Real Rape*. Harvard University Press.

Falconer, D. 1981. *Introduction to Quantitative Genetics*. Longman.

Farr, J. 1980. The effects of sexual experience and female receptivity on courtship-rape decisions in male guppies, *Poecilla reticulate*. *Animal Behaviour* 29: 1272–1273.

Farr, J., J. Travis, and J. Trexler. 1986. Behavioral allometry and interdemic variation in sexual behavior of the sailfin molly *Poecilla latipinna* Pisces Poecilidae. *Animal Behaviour* 34: 497–509.

Fausto-Sterling, A. 1985. *Myths of Gender: Biological Theories about Women and Men*. Basic Books.

Fausto-Sterling, A. 1997. Feminism and behavioral evolution: A taxonomy. In *Feminism and Evolutionary Biology*, ed. P. Gowaty. Chapman and Hall.

Felson, R., and M. Krohn. 1990. Motives for rape. *Journal of Research in Crime and Delinquency* 27: 222–242.

Feshbach, N. 1969. Sex differences in children's modes of aggressive responses toward outsiders. *Merrill-Palmer Quarterly* 15: 249–258.

Field, S. 1998. Of souls and skyhooks. *Trends in Ecology and Evolution* 13: 296.

Figueredo, A., and L. McCloskey. 1993. Sex, money and paternity: The evolutionary psychology of domestic violence. *Ethology and Sociobiology* 14: 353–379.

Figueredo, A. J., B. D. Sales, J. V. Becker, K. Russell, and M. Kaplan. 1999. A Brunswikian evolutionary-developmental model of adolescent sex offending. *Behavioral Sciences and the Law* (in press).

Finkelhor, D., and K. Yllo. 1985. *License to Rape: Sexual Abuse of Wives*. Holt, Rinehart and Winston.

Fishelson, L. 1970. Behaviour and ecology of a population of *Abudefdluf saxatalis* (Oinacebtriidae: Teleostei) at Eliat (Red Sea). *Animal Behaviour* 18: 225–237.

Fisher, S. 1973. *The Female Orgasm: Psychology, Physiology and Fantasy*. Basic Books.

Flinn, M. 1987. Mate guarding in a Caribbean village. *Ethology and Sociobiology* 8: 1–28.

Flinn, M. 1988. Parent-offspring interactions in a Caribbean village: Daughter guarding. In *Human Reproductive Behavior*, ed. L. Betzig et al. Cambridge University Press.

Flinn, M. 1997. Culture and the evolution of social learning. *Evolution and Human Behavior* 18: 23–67.

Flinn, M., and R. Alexander. 1982. Culture theory: The developing synthesis from biology. *Human Ecology* 10: 383–400.

Fonow, M., L. Richardson, and V. Wemmerus. 1992. Feminist rape education: Does it work? *Gender and Society* 6: 108–121.

Ford, C., and F. Beach. 1951. *Patterns of Sexual Behavior*. Harper and Row.

Forman, S. 1967. Cognition and the catch: The location of fishing spots in a Brazilian coastal village. *Ethnology* 6: 417–426.

Freeman, D. 1983. *Margaret Mead and Samoa: The Making and Unmaking of an Anthropological Myth*. Harvard University Press.

Freud, S. 1933. *New Introductory Lectures on Psychoanalysis*. Norton.

Fuller, P. 1995. The social construction of rape in appeal court cases. *Feminism and Psychology* 5: 154–161.

Furlow, B., S. Gangestad, and T. Armijo-Pruett. 1998. Developmental stability and human violence. *Proceedings of the Royal Society of London* B 265: 1–6.

Furnham, A., and N. Brown. 1996. Theories of rape and the just world. *Psychology, Crime and Law* 2: 211–229.

Galdikas, B. 1979. Orangutan adaptation at Tanjung-Putting Reserve: Mating and ecology. In *The Great Apes*, ed. D. Hamburg and E. McCown. Benjamin/Cummings.

Galdikas, B. 1985a. Adult male sociality and reproductive tactics among orangutans at Tanjung-Putting Borneo Indonesia. *Folia Primatology* 45: 9–24.

Galdikas, B. 1985b. Subadult male orangutan sociality and reproductive behavior at Tanjung-Putting. *American Journal of Primatology* 8: 87–100.

Galdikas, B. 1995. *Reflections of Eden: My Years with the Orangutans of Borneo*. Little, Brown.

Gangestad, S. 1993. Sexual selection and physical attractiveness: Implications for mating dynamics. *Human Nature* 4: 205–236.

Gangestad, S. 1997. Evolutionary psychology and genetic variation: Non-adaptative, fitness related and adaptive. In *Characterizing Human Psychological Adaptations*, ed. G. Bock and G. Cardew. Wiley.

Gangestad, S., and R. Thornhill. 1997a. Human sexual selection and developmental stability. In *Evolutionary Social Psychology*, ed. J. Simpson and D. Kendrick. Erlbaum.

Gangestad, S., and R. Thornhill. 1997b. The evolutionary psychology of extra-pair sex: The role of fluctuating asymmetry. *Evolution and Human Behavior* 18: 69–88.

Gangestad, S., and R. Thornhill. 1998. Menstrual cycle variation in women's preference for the scent of symmetrical men. *Proceedings of the Royal Society of London* B 265: 927–933.

Gangestad, S., and R. Thornhill. 1999. Individual differences in developmental precision and fluctuating asymmetry: A model and its implications. *Journal of Evolutionary Biology* 12: 402–416.

Gazzaniga, M. 1989. Organization of the human brain. *Science* 245: 947–952.

Gazzaniga, M., ed. 1995. *The Cognitive Neurosciences*. MIT Press.

Geary, D. 1998. *Male, Female: The Evolution of Human Sex Differences*. American Psychological Association.

Geary, D., M. Rumsey, C. Bow-Thomas, and K. Hoard. 1995. Sexual jealousy as a facultative trait: Evidence from the pattern of sex differences in adults from China and the United States. *Ethology and Sociobiology* 16: 355–383.

Gebhard, P., C. Christenson, J. Gagnon, and W. Pomeroy. 1965. *Sex Offenders: An Analysis of Types*. Harper and Row.

Geis, G. 1977. Forcible rape: An introduction. In *Forcible Rape*, ed. D. Chappell et al. Columbia University Press.

Ghiglieri, M. 1999. *The Dark Side of Man*. Perseus.

Gill, S. 1996. Dismantling gender and race stereotypes: Using education to prevent date rape. *UCLA Women's Law Journal* 7: 27–79.

Gladstone, D. 1979. Promiscuity in monogamous colonial birds. *American Naturalist* 114: 545–559.

Goethals, G. 1971. Biological influences on sexual identify. In *Human Sexuality*, ed. H. Katchadourian. Basic Books.

Goldfarb, C. 1984. Practice of using castration in sentence being questioned. *Criminal Justice Newsletter* 15 (February 15): 3–4.

Goodall, J. 1986. *The Chimpanzees of Gombe: Patterns of Behavior*. Harvard University Press.

Gould, S. 1987. *An Urchin in the Storm: Essays about Books and Ideas*. Norton.

Gould, S. 1997. Darwinian fundamentalism. *New York Review of Books*, June 12, p. 26.

Gould, S., and R. Lewontin. 1979. The spandrels of San Marco and the Panglossian paradigm: A critique of the adaptationist program. *Proceedings of the Royal Society of London* B 205: 581–598.

Gowaty, P. 1982. Sexual terms in sociobiology: Emotionally evocative and paradoxically, jargon. *Animal Behaviour* 30: 630–631.

Gowaty, P. 1992. Evolutionary biology and feminism. *Human Nature* 3: 217–249.

Gowaty, P., ed. 1997. *Feminism and Evolutionary Biology: Boundaries, Intersections, and Frontiers*. Chapman and Hall.

Gowaty, P., and N. Buschhaus. 1998. Ultimate causation of aggressive and forced copulation in birds: Female resistance, the CODE hypothesis, and social monogamy. *American Zoologist* 38: 207–225.

Gowaty, P., and D. Mock, eds. 1985. Avian monogamy. Ornithological Monographs, No. 37. American Ornithologists' Union.

Grammer, K. 1993. Signale der Liebe die Biologischen gesetz der partnerschaft. Hoffman und Campe.

Grammer, K., and R. Thornhill. 1994. Human (*Homo sapiens*) facial attractiveness and sexual selection: The role of symmetry and averageness. *Journal of Comparative Psychology* 108: 233–242.

Grano, J. 1990. Free speech v. The University of Michigan. *Academic Questions*, Spring: 7–22.

Gray, R. 1997. "In the belly of the monster": Feminism, developmental systems, and evolutionary explanations. In *Feminism and Evolutionary Biology*, ed. P. Gowaty. Chapman and Hall.

Green, J., R. Copelan, P. Cotter, and B. Stephens. 1994. Affecting the rules for the prosecution of rape and other gender-based violence before the international criminal tribunal for the former Yugoslavia: A feminist proposal and critique. *Hasting's Women's Law Journal* 5: 171–240.

Greenfield, L. 1997. Sex Offenses and Offenders: An Analysis of Data on Rape and Sexual Assault. Bureau of Justice Statistics, US Department of Justice.

Greenfield, L., M. Rand, D. Graven, P. Klaus, C. Perkins, C. Ringel, G. Warchol, and C. Maston. 1998. Violence By Intimates: Analysis of Data On Crimes By Current Or Former Spouses, Boyfriends, and Girlfriends. Bureau of Justice Statistics, US Department of Justice.

Greer, G. 1970. *The Female Eunuch*. Bantam.

Griffin, S. 1971. Rape: The all-American crime. *Ramparts* 10: 26–36.

Gross, P., and N. Levin. 1994. *Higher Superstition: The Academic Left and Its Quarrels with Science*. Johns Hopkins University Press.

Groth, N. 1979. *Men Who Rape*. Plenum.

Groth, N., and H. Birnbaum. 1986. The rapist: Motivations for sexual violence. In *The Rape Crisis Center Handbook*, ed. S. McCombie. Plenum.

Groth, N., and W. Hobson. 1983. The dynamics of sexual thought. In *Sexual Dynamics of Anti-Social Behavior*, ed. L. Schelsinger and E. Revitch. Thomas.

Hagen, R. 1979. *The Biosexual Factor*. Doubleday.

Hall, G., D. Shondrick, and R. Hirschman. 1993. The role of sexual arousal in sexually aggressive behavior: A meta-analysis. *Journal of Consulting and Clinical Psychology* 61: 1091–1095.

Hamilton, W. 1963. The evolution of altruistic behavior. *American Naturalist* 97: 354–356.

Hamilton, W. 1964. The genetical evolution of social behavior, parts 1 and 2. *Journal of Theoretical Biology* 7: 1–52.

Hamilton, W. 1966. The moulding of senescence by natural selection. *Journal of Theoretical Biology* 12: 12–45.

Hammond, H., J. Redman, and C. Caskey. 1995. *In utero* paternity testing following alleged sexual assault. *Journal of the American Medical Association* 273: 1774–1777.

Harding, C. 1985. Sociobiological hypotheses about rape: A critical look at the data behind the hypotheses. In *Violence against Women*, ed. S. Sunday and E. Tobach. Gordian Press.

Harris, G., and M. Rice. 1996. The science in phallometric testing of men's sexual preferences. *Current Directions in Psychological Science* 5: 156–160.

Harris, M. 1989. *Our Kind*. Harper Collins.

Hartung, J. 1992. Getting real about rape. *Behavioral and Brain Sciences* 15: 390–392.

Hempel, C. 1959. The logic of functional analysis. In *Symposium on Sociological Theory*, ed. L. Gross. Harper and Row.

Hemni, Y., T. Koga, and M. Murai. 1993. Mating behavior of the sand bubbler crab, *Scopimera globosa*. *Journal of Crustacean Biology* 13: 736–744.

Herman, J. 1990. Sex offenders: A feminist perspective. In *Handbook of Sexual Assault*, ed. W. Marshall et al. Plenum.

Hewlett, B., ed. 1992. *Father-Child Relations: Cultural and Biosocial Contexts*. Aldine de Gruyter.

Hicks, P. 1993. Comment: Castration of sexual offenders, legal and ethical issues. *Journal of Legal Medicine* 14: 641–644.

Hill, K., and A. Hurtado. 1996. *Ache Life History: The Ecology and Demography of a Foraging People*. Aldine de Gruyter.

Hilton, D. 1982. Is it really rape or forced copulation? *Bioscience* 32: 641.

Hindelang, M. 1977. *Criminal Victimization in Eight American Cities: A Descriptive Analysis of Common Theft and Assault*. Ballinger.

Hindelang, M., and B. Davis. 1997. Forcible rape in the United States: A statistical profile. In *Forcible Rape*, ed. D. Chappell et al. Columbia University Press.

Holmes, M., H. Resnick, D. Kilpatrick, and C. Best. 1996. Rape-related pregnancy: Estimates and descriptive characteristics from a national sample of women. *American Journal of Obstetrics and Gynecology* 175: 320–325.

Holton, G. 1993. *Science and Anti-Science*. Harvard University Press.

Holton, G. 1996. *Einstein, History, and Other Passions: The Rebellion against Science at the End of the Twentieth Century*. Addison-Wesley.

Hoogland, J., and P. Sherman. 1976. Advantages and disadvantages of bank swallow coloniality. *Ecological Monographs* 46: 33–58.

Howard, R. 1978. The evolution of mating strategies in bull frogs, *Rana catesbeiana*. *Evolution* 32: 850–871.

Humphrey, N. 1980. Nature's psychologists. In *Consciousness and the Physical World*, ed. B. Josephson and V. Ramachandran. Pergamon.

Hursch, C. 1977. *The Trouble with Rape*. Nelson-Hall.

Icenogle, D. 1994. Sentencing male sex offenders to the use of biological treatments: A Constitutional analysis. *Journal of Legal Medicine* 15: 279–304.

Jackson, S. 1995. The social context of rape: Sexual scripts and motivation. In *Rape and Society*, ed. P. Searles and R. Berger. Westview.

Jacobson, N., and J. Gottman. 1998. *When Men Batter Women*. Simon and Schuster.

Johnston, V., and M. Franklin. 1993. Is beauty in the eye of the beholder? *Ethology and Sociobiology* 14: 183–199.

Jones, A. 1990. Family matters. In *The Sexual Liberals and the Attack on Feminism*, ed. D. Leidholdt and J. Raymond. Pergamon.

Jones, C. 1985. Reproductive patterns in mantled howler monkeys estrus mate choice and copulation. *Primates* 26: 130–142.

Jones, D. 1996. Physical Attractiveness and the Theory of Sexual Selection: Results From Five Populations. Museum of Anthropology, University of Michigan, Ann Arbor.

Jones, O. 1999. Sex, culture and the biology of rape: Toward explanation and prevention. *California Law Review* 87: 827–942.

Kacelnik, A. 1997. Normative and descriptive models of decision making: Time discounting and risk sensitivity. In *Characterizing Human Psychological Adaptations*, ed. G. Bock and G. Cardew. Wiley.

Kalichman, S., E. Williams, C. Cherry, L. Belcher, and D. Nachimson. 1998. Sexual coercion, domestic violence, and negotiating condom use among low-income African-American women. *Journal of Women's Health* 7: 371–378.

Kanin, E. 1994. False rape allegations. *Archives of Sexual Behavior* 23: 81–90.

Katz, S., and M. Mazur. 1979. *Understanding the Rape Victim*. Wiley.

Keeneyside, M. 1972. Intraspecific intrusions into nests of spawning longear sunfish. *Copeia* 272–278.

Kenrick, D. 1989. Bridging social psychology and sociobiology: The case of sexual attraction. In *Sociobiology and the Social Sciences*, ed. R. Bell and N. Bell. Texas Tech University Press.

Kenrick, D., M. Trost, and V. Sheets. 1996. Power, harassment, and trophy mates: The feminist advantages of an evolutionary perspective. In *Sex, Power, Conflict*, ed. D. Buss and N. Malamuth. Oxford University Press.

Kilpatrick, D., C. Edmunds, and A. Seymour. 1992. Rape in America: A Report to the Nation. National Victim Center, Arlington, Virginia.

Kilpatrick, D., B. Saunders, C. Best, and J. Von. 1987. Criminal victimization: Lifetime prevalence, reporting to police, and psychological impact. *Crime and Delinquency* 33: 479–489.

Kinsey, A., W. Pomeroy, and C. Martin. 1948. *Sexual Behavior in the Human Male*. Saunders.

Kitcher, P. 1985. *Vaulting Ambition: Sociobiology and the Quest for Human Nature*. MIT Press.

Kodric-Brown, A. 1977. Reproductive success and the evolution of breeding territories in pupfish (*Cyprinodon*). *Evolution* 31: 750–766.

Kopp, M. 1938. Surgical treatment as sex crime prevention measure. *Journal of Criminal Law, Criminology, and Police Science* 28: 692–706.

Koss, M., C. Gidycs, and N. Wisniewski. 1987. The scope of rape: Incidence and prevalence of sexual aggression and victimization in a national sample of higher education students. *Journal of Consulting and Clinical Psychology* 55: 162–170.

Kramer, L. 1987. Albuquerque Rape Crisis Center: Annual Report. Bernalillo County Mental Health/Mental Retardation Center.

Krebs, J., and N. Davies. 1993. *An Introduction to Behavioral Ecology*, third edition. Blackwell.

Lalumiére, M., and V. Quinsey. 1994. The discriminability of rapists from non-sex offenders using phallometric measures: A meta-analysis. *Criminal Justice and Behavior* 21: 150–175.

Lalumiére, M., and M. Seto. 1998. What's wrong with psychopaths? Defining the causes and effects of psychopathy. *Psychiatry Rounds* 2 (6).

Lalumiére, M., L. Chalmers, V. Quinsey, and M. Seto. 1996. A test of the mate deprivation hypothesis of sexual coercion. *Ethology and Sociobiology* 17: 299–318.

Lancaster, J. 1997. The evolutionary history of human parental investment in relation to population growth and social stratification. In *Feminism and Evolutionary Biology*, ed. P. Gowaty. Chapman and Hall.

Langan, P., and C. Harlow. 1994. Child Rape Victims, 1992. Crime Data Brief, Bureau of Justice Statistics, US Department of Justice.

Las, A. 1972. Male courtship persistence in the greenhouse whitefly, *Trialeurodae vaoporarior* um Westwook (Homoptera: Aleyrodidae). *Behaviour* 72: 107–126.

Lederer, L. 1980. *Playboy* isn't playing: An interview with Judith Bat-Ada. In *Take Back the Night*, ed. L. Lederer. William Morrow.

LeGrand, C. 1973. Rape and rape laws: Sexism in society and law. *California Law Review* 8: 263–294.

Leslie, C. 1990. Scientific racism: Reflections on peer review, science and ideology. *Social Science and Medicine* 31: 891–912.

LeVine, R. 1977. Gusii sex offenses: A study in social control. In *Forcible Rape*, ed. D. Chappell et al. Columbia University Press.

Lewontin, R., S. Rose, and L. Kamin. 1984. *Not in Our Genes: Biology, Ideology and Human Nature*. Pantheon.

Linley, J. 1972. A study of the mating behavior of *Colicoides melleus* (Copuillet) (Diptera: Ceratopogonidae). *Transactions of the Royal Entomological Society of London* 126: 279–303.

Littlewood, R. 1997. Military rape. *Anthropology Today* 13: 7–16.

Lohr, B., H. Adams, and J. Davis. 1997. Sexual arousal to erotic and aggressive stimuli in sexually coercive and noncoercive men. *Journal of Abnormal Psychology* 106: 230–242.

Lonsway, K., and L. Fitzgerald. 1994. Rape myths: In review. *Psychology of Women Quarterly* 18: 133–164.

Lorenz, K. 1970. *Studies in Animal and Human Behavior*, volume 1. Harvard University Press.

Lorenz, K. 1971. *Studies in Animal and Human Behavior*, volume 2. Harvard University Press.

Losco, J. 1981. Ultimate vs. proximate explanation: Explanation modes in sociobiology and the social sciences. *Journal of Social and Biological Structures* 4: 329–346.

Low, B. 1978. Environmental uncertainty and the parental strategies of marsupials and placentals. *American Naturalist* 112: 197–213.

Low, B. 1989. Cross-cultural patterns in the training of children: An evolutionary perspective. *Journal of Comparative Psychology* 103: 311–319.

Lykken, D. 1995. *The Antisocial Personalities*. Erlbaum.

MacDonald, J. 1971. *Rape Offenders and Their Victims*. Thomas.

MacDonald, K., ed. 1988. *Sociobiological Perspectives on Human Development*. Springer-Verlag.

MacDonald, K. 1992. Warmth as a developmental construct: An evolutionary analysis. *Child Development* 63: 753–774.

MacKinnon, C. 1983. Feminism, Marxism, method and the state: Toward feminist jurisprudence. *Signs* 8 (4): 635–658.

MacKinnon, C. 1987. *Feminism Unmodified: Discourses on Life and Law*. Harvard University Press.

MacKinnon, C. 1989. *Toward a Feminist Theory of State*. Harvard University Press.

MacKinnon, C. 1990. Liberalism and the death of feminism. In *The Sexual Liberals and the Attack on Feminism*, ed. D. Leidholdt and J. Raymond. Pergamon.

MacKinnon, C. 1993. *Only Words*. Harvard University Press.

MacKinnon, J. 1974. *In Search of the Red Ape*. Holt, Rinehart, and Winston.

MacKinnon, J. 1979. Reproductive behavior in wild orangutan populations. In *The Great Apes*, ed. D. Hamburg and E. McCown. Benjamin/Cummings.

Malamuth, N. 1989. The attraction to sexual aggression scale: Part two. *Journal of Sex Research* 26: 324–354.

Malamuth, N. 1996. The confluence model of sexual aggression: Feminist and evolutionary perspectives. In *Sex, Power, Conflict*, ed. D. Buss and N. Malamuth. Oxford University Press.

Malamuth, N. 1998. An evolutionary-based model integrating research on the characteristics of sexually coercive men. In *Advances in Psychological Science*, volume 2: *Personal, Social, and Developmental Aspects*, ed. J. Adair et al. Psychology Press.

Malamuth, N., and M. Heilmann. 1998. Evolutionary psychology and sexual aggression. In *Handbook of Evolutionary Psychology*, ed. C. Crawford and D. Krebs. Erlbaum.

Malamuth, N., and D. Linz. 1993. *Pornography: Social Aspects*. Sage.

Malinowski, B. 1929. *The Sexual Life of Savages in North-Western Melanesia*. Halcyon House.

Manning, A. 1967. The control of sexual receptivity in female *Drosophila*. *Animal Behaviour* 15: 239–250.

Manning, J., D. Scutt, G. Whitehouse, and S. Leinster. 1997. Breast asymmetry and phenotypic quality in women. *Evolution and Human Behavior* 18: 223–236.

Marr, D. 1982. *Vision*. Freeman.

Marshall, W., and S. Barrett. 1990. *Criminal Neglect: Why Sex Offenders Go Free*. McClelland-Banton.

Masters, R., B. Hone, and A. Doshi. 1998. Environmental pollution, neurotoxicity, and criminal violence. In *Environmental Toxicology*, ed. J. Rose. Gordon and Breach.

Maynard Smith, J. 1995. *New York Review of Books*, November 30, p. 46.

Maynard Smith, J. 1997. Commentary. In *Feminism and Evolutionary Biology: Boundaries, Intersections, and Frontiers*, ed. P. Gowaty. Chapman and Hall.

Maynard Smith, J. 1998. The origin of altruism. *Nature* 393: 639–640.

Mayr, E. 1983. How to carry out the adaptationist program? *American Naturalist* 121: 324–334.

McCahill, T., L. Meyer, and A. Fischman. 1979. *The Aftermath of Rape*. Heath.

McCay, B. 1978. Systems ecology, people ecology and the anthropology of fishing communities. *Human Ecology* 6: 397–422.

McKinley, J., Jr. 1996. Rwanda's war legacy: Children born of rape. *New York Times*, September 13.

McKinney, F., and P. Stolen. 1982. Extra-pair-courtship and forced copulation among captive green-winged teal (*Anas crecca carolinensis*). *Animal Behaviour* 30: 461–474.

McKinney, F., J. Barret, and S. Derrickson. 1980. Rape among mallards. *Science* 281–282.

McKinney, F., S. Derrickson, and P. Mineau. 1983. Forced copulation in waterfowl. *Behaviour* 86: 250–294.

Mead, M. 1935. *Sex and Temperament in Three Primitive Societies*. Dell.

Mealey, L. 1995. The sociobiology of sociopathy: An integrated evolutionary model. *Behavioral and Brain Sciences* 18: 523–541.

Mealey, L., R. Bridgstock, and G. Townsend. 1999. Symmetry and perceived facial attractiveness: A monozygotic co-twin comparison. *Journal of Personality and Social Psychology* 76: 151–158.

Mesnick, S. 1997. Sexual alliances: Evidence and evolutionary implications. In *Feminism and Evolutionary Biology*, ed. P. Gowaty. Chapman and Hall.

Mesnick, S., and B. Le Boeuf. 1991. Sexual behavior of male northern elephant seals. 2. Female response to potentially injurious encounters. *Behaviour* 117: 262–280.

Miccio, K. 1994. Rape is a gender based crime. In *Crimes of Gender: Violence against Women*, ed. G. McCuen. McCuen.

Millett, K. 1971. *The Prostitution Papers: A Candid Dialogue*. Basic Books.

Mineau, P., and F. Cooke. 1979. Rape in the lesser snow goose. *Behaviour* 70: 280–291.

Minturn, L., M. Grosse, and S. Haider. 1969. Cultural patterning of sexual beliefs and behavior. *Ethnology* 8: 301–318.

Mitani, J. 1985. Mating behaviour of male orangutans in the Kutai game reserve, Indonesia. *Animal Behaviour* 33: 392–402.

Miyazawa, K. 1976. Victimological studies of sexual crimes in Japan. *Victimology* 1: 107–129.

Møller, A., and R. Alatalo. 1999. Good genes effects in sexual selection. *Proceedings of the Royal Society of London* B 266: 85–91.

Møller, A., and J. Swaddle. 1997. *Asymmetry, Developmental Stability, and Evolution*. Oxford University Press.

Møller, A., and R. Thornhill. 1997. A meta-analysis of the heritability of developmental stability. *Journal of Evolutionary Biology* 10: 1–16.

Møller, A., and R. Thornhill. 1998a. Bilateral symmetry and sexual selection: A meta-analysis. *American Naturalist* 151: 174–192.

Møller, A., and R. Thornhill. 1998b. Male parental care, differential parental investment by females, and sexual selection. *Animal Behaviour* 55: 1507–1515.

Moore, G. 1903. *Principia Ethica*. Cambridge University Press.

Moore, J. 1996. Discussion of Holmes et al. 1996. *American Journal of Obstetrics and Gynecology* 175: 324–325.

Morgan, J. 1981. Relationship between rape and physical damage during rape and phase of sexual cycle during which rape occurred. Doctoral dissertation, University of Texas at Austin.

Morris, A. 1987. *Women, Crime, and Criminal Justice*. Blackwell.

Morris, M. 1996. By force of arms: Rape, war and military culture. *Duke Law Journal* 45: 651–771.

Morris, N., and J. Udry. 1970. Variations in pedometer activity during the menstrual cycle. *Obstetrics and Gynecology* 35: 199–201.

Morrison, C. 1986. A cultural perspective on rape. In *The Rape Crisis Intervention Handbook*, ed. S. McCombie. Plenum.

Muehlenhard, C., S. Danoff-Burgg, and I. Powch. 1996. Is rape sex or violence? Conceptual issues and implications. In *Sex, Power, Conflict*, ed. D. Buss and N. Malamuth. Oxford University Press.

Murdock, G. 1972. Anthropology's mythology. In Proceedings of the Royal Anthropological Institute of Great Britain and Ireland for 1971.

Murphey, D. 1992. Feminism and rape. *Journal of Social, Political and Economic Studies* 17: 13–27.

Mynatt, C., and E. Allgeier. 1990. Risk factor, self-attributions, and adjustment problems among victims of sexual coercion. *Journal of Applied Social Psychology* 20: 130–153.

Nadler, R. 1999. Sexual aggression in the great apes: Implications for human law. *Jurimetrics* 39: 149–155.

Nadler, R., and L. Miller. 1982. Influence of male aggression on mating of gorillas. *Folia Primatologica* 38: 233–239.

Niarcos, C. 1995. Women, war and rape: Challenges facing the international tribunal for the former Yugoslavia. *Human Rights Quarterly* 17: 649–690.

Nesse, R., and G. Williams. 1994. *Why We Get Sick: The New Science of Darwinian Medicine*. Times Books.

Nowak, M., and K. Sigmund. 1998. Evolution of indirect reciprocity by image scoring. *Nature* 393: 573–577.

Nunney, L. 1998. Are we selfish, are we nice, or are we nice because we are selfish? *Science* 281: 1619–1621.

Oh, R. 1979. Repeated copulation in the brown planthopper, *Nilaparvata lugens* Stal (Homo pterea: Delphacidae). *Ecological Entomology* 4: 345–353.

Oyama, S. 1985. *The Ontogeny of Information: Developmental Systems and Evolution*. Cambridge University Press.

Paglia, C. 1992. *Sex, Art, and American Culture*. Vintage.

Paglia, C. 1994. *Vamps and Tramps*. Vintage.

Paglia, C. 1996. Quoted in *Newsweek*, June 3, p. 69.

Palmer, C. 1988a. Twelve reasons why rape is not sexually motivated: A skeptical examination. *Journal of Sex Research* 25 (4): 512–530.

Palmer, C. 1988b. Evolutionary explanations of rape: Testing adaptive and nonadaptive hypotheses. Doctoral dissertation, Arizona State University.

Palmer, C. 1989a. Is rape a cultural universal? A re-examination of the ethnographic evidence. *Ethnology* 28: 1–16.

Palmer, C. 1989b. Rape in nonhuman species: Definitions, evidence, and implications. *Journal of Sex Research* 26: 353–374.

Palmer, C. 1991. Human rape: Adaptation or by-product? *Journal of Sex Research* 28: 365–386.

Palmer, C. 1992a. The use and abuse of Darwinian psychology: Its impact on attempts to determine the evolutionary basis of human rape. *Ethology and Sociobiology* 13: 289–299.

Palmer, C. 1992b. Behavior vs. psychological mechanisms: Does the difference really make a difference? *Behavioral and Brain Sciences* 15: 402–403.

Palmer, C. 1993. Anger, aggression, and humor in Newfoundland floor hockey: An evolutionary analysis. *Aggressive Behavior* 19: 167–173.

Palmer, C. 1994. Folk management, 'soft evolutionism,' and fishers' motives: Implications for the regulation of the lobster fisheries of Maine and Newfoundland. *Human Organization* 52: 414–420.

Palmer, C., and L. Steadman. 1997. Human kinship as a descendant-leaving strategy: A solution to an evolutionary puzzle. *Journal of Social and Evolutionary Systems* 20: 39–51.

Palmer, C., D. DiBari, and S. Wright. 1999. Is it sex yet? Theoretical and practical implications of the debate over rapists' motives. *Jurimetrics* 39: 271–282.

Palmer, C., B. Fredrickson, and C. Tilley. 1997. Categories and gatherings: Group selection and the mythology of cultural anthropology. *Evolution and Human Behavior* 18: 291–308.

Parker, G. 1970. Sperm competition and its evolutionary consequences in the insects. *Biological Reviews* 45: 525–568.

Parker, G. 1974. Courtship persistence and female guarding as male time-investment strategies. *Behaviour* 48: 157–184.

Parker, G. 1979. Sexual selection and sexual conflict. In *Sexual Selection and Reproductive Competition in Insects*, ed. M. Blum and N. Blum. Academic Press.

Parrot, A. 1991. Vital childhood lessons: The role of parenting in preventing sexual coercion. In *Sexual Coercion*, ed. E. Grauerholz and M. Koralewski. Lexington.

Pawson, E., and G. Banks. 1993. Rape and fear in a New Zealand city. *Area* 25: 55–63.

Penn, D., and W. Potts. 1998. Chemical signals and parasite-mediated sexual selection. *Trends in Ecology and Evolution* 13: 391–396.

Penton-Voak, I., D. Perrett, D. Burt, D. Castles, T. Koyabashi, L. Murray, and R. Minamisawa. 1999. Female preference for male faces changes cyclically. *Nature* 399: 741–742.

Perkins, C., and P. Klaus. 1996. Criminal Victimization 1994. National Crime Victimization Survey. Bulletin, Bureau of Justice Statistics, US Department of Justice.

Perkins, C., P. Klaus, L. Bastian, and R. Cohen. 1996. Criminal Victimization in the United States, 1993. National Crime Victimization Survey Report. Bureau of Justice Statistics, US Department of Justice.

Perrett, D., K. Lee, I. Penton-Voak, D. Rowland, S. Yoshikawa, D. Burt, S. Henzi, D. Castles, and S. Kamatsu. 1998. Effects of sexual dimorphism on facial attractiveness. *Nature* 394: 884–887.

Perusse, D. 1993. Cultural and reproductive success in industrial societies: Testing the relationship at the ultimate and proximate levels. *Behavioral and Brain Sciences* 16: 267–322.

Pinker, S. 1994. *The Language Instinct: How the Mind Creates Language*. William Morrow.

Pinker, S. 1997. *How the Mind Works*. Norton.

Pinkser, W., and E. Doschek. 1980. Courtship and rape: The mating behaviour of *Drosophila subobscura* in light and darkness. *Zeitschrift für Tierpsychologie* 54: 57–70.

Pinto, J. 1972. A synopsis of the bionomics of *Phoda alticeps* (Coleoptera: Meloidae) with special reference to sexual behavior. *Canadian Entomologist* 104: 577–595.

Podhoretz, N. 1991. Rape in feminist eyes. *Commentary* 11: 29–35.

Polaschek, D., T. Ward, and S. Hudson. 1997. Rape and rapists: Theory and treatment. *Clinical Psychology Review* 17: 117–144.

Popper, K. 1968. *Conjectures and Refutations*. Harper and Row.

Provine, W. 1971. *The Origins of Theoretical Population Genetics*. University of Chicago Press.

Proulx, J., J. Aubut, A. McKibben, and M. Côte. 1994. Penile responses of rapists and nonrapists to rape stimuli involving physical violence and humiliation. *Archives of Sexual Behavior* 23: 295–310.

Pulliam, H., and C. Dunford. 1980. *Programmed to Learn: An Essay on the Evolution of Culture*. Columbia University Press.

Queen's Bench Foundation. 1978. The rapist and his victim. In *Crime and Society*, ed. L. Savitz and N. Johnston. Wiley.

Queller, D. 1995. The spandrels of Saint Marx and the panglossian paradox: A critique of a rhetorical programme. *Quarterly Review of Biology* 70: 485–489.

Quinsey, V., and M. Lalumiére. 1995. Evolutionary perspectives on sexual offending. *Sexual Abuse* 7: 301–315.

Quinsey, V., T. Chaplin, and D. Upfold. 1984. Sexual arousal to nonsexual violence and sadomasochistic themes among rapists and non-sex-offenders. *Journal of Consulting and Clinical Psychology* 52: 651–657.

Quinsey, V., M. Rice, G. Harris, and K. Reid. 1993. The phylogenetic and ontogenetic development of sexual age preferences in males: Conceptual and measurement issues. In *The Juvenile Sex Offender*, ed. H. Barbaree et al. Guilford.

Rada, R., ed. 1978a. *Clinical Aspects of the Rapist*. Grune and Stratton.

Rada, R. 1978b. Psychological factors in rapist behavior. In *Clinical Aspects of the Rapist*, ed. R. Rada. Grune and Stratton.

Rada, R. 1978c. Biological aspects and organic treatment of the rapist. In *Clinical Aspects of the Rapist*, ed. R. Rada. Grune and Stratton.

Rappaport, R. 1967. *Pigs for the Ancestors: Ritual in the Ecology of a New Guinea People*. Yale University Press.

Raymond, J. 1982. Rhetoric: The methodology of the humanities. *College English* 44: 778–783.

Resick, P. 1993. The psychological impact of rape. *Journal of Interpersonal Violence* 8: 223–255.

Resnick, H., D. Kilpatrick, B. Dansky, B. Saunders, and C. Best. 1993. Prevalence of civilian trauma and posttraumatic stress syndrome in a representative national sample of women. *Journal of Consulting and Clinical Psychology* 61: 984–991.

Ridley, M., ed. 1987. *The Darwin Reader*. Norton.

Ridley, M. 1993. *The Red Queen: Sex and the Evolution of Human Nature*. Macmillan.

Riger, S., and M. Gordon. 1981. The fear of rape: A study in social control. *Journal of Social Issues* 37: 71–89.

Rijksen, H. 1978. *A Field Study on Sumatran Orangutans*. Veenan and Zonen.

Rodabaugh, B., and M. Austin. 1981. *Sexual Assault: A Guide for Community Action*. Garland STPM.

Rogel, M. 1976. Biosocial aspects of rape. Doctoral dissertation, University of Chicago.

Rogers, P. 1995. Male rape: The impact of a legal definition on the clinical area. *Medical Science and Law* 35: 303–306.

Rose, S. 1998. *Lifelines*. Oxford University Press.

Rosenqvist, G. 1990. Male mate choice and female-female competition for mates in the pipefish *Nerophis ophidion*. *Animal Behaviour* 39: 1110–1116.

Rowe, L., G. Arnqvist, A. Sih, and J. Krupa. 1994. Sexual conflict and the evolutionary ecology of mating patterns: Water striders as a model system. *Trends in Ecology and Evolution* 9: 289–293.

Rozée, D. 1993. Forbidden or forgiven? Rape in cross-cultural perspective. *Psychology of Women Quarterly* 17: 499–514.

Rubenstein, D., and R. Wrangham. 1986. *Ecological Aspects of Social Evolution: Birds and Mammals*. Princeton University Press.

Ruse, M. 1994. Struggle for the soul of science. *The Sciences* 34: 39–44.

Russell, D. 1975. *The Politics of Rape: The Victim's Perspective*. Stein and Day.

Russell, D. 1982. *Rape in Marriage*. Macmillan.

Russell, D. 1984. *Sexual Exploitation: Rape, Child Sexual Abuse and Workplace Harassment*. Sage.

Sahlins, M. 1976. *The Use and Abuse of Biology: An Anthropological Critique of Sociobiology*. University of Michigan Press.

Sakaluk, S., P. Bangert, A.-K. Eggert, C. Gack, and L. Swanson. 1995. The gin trap as a device facilitating coercive mating in sagebrush crickets. *Proceedings of the Royal Society of London* B 261: 65–71.

Sanday, P. 1981. The socio-cultural context of rape: A cross-cultural study. *Journal of Social Issues* 37 (4): 5–27.

Sanday, P. 1990. *Fraternity Gang Rape*. New York University Press.

Sanders, W. 1980. *Rape and Women's Identity*. Sage.

Sanford, L., and A. Fetter. 1979. *In Defense of Ourselves*. Doubleday.

Schiff, A. 1971. Rape and other countries. *Medicine, Science, and Law* 11: 139–143.

Schwagmeyer, P. 1988. Scramble-competition polygyny in an asocial mammal: Male mobility and mating success. *American Naturalist* 131: 885–892.

Schwendinger, J., and J. Schwendinger. 1985. *Homo economicus* as rapist. In *Violence against Women*, ed. S. Sunday and E. Tobach. Gordian Press.

Scott, E. 1993. How to stop rapists? A question of strategy in two rape crisis centers. *Social Problems* 40: 343–361.

Scully, D., and J. Marolla. 1995. Riding the bull at Gilley's: Convicted rapists describe the rewards of rape. In *Rape and Society*, ed. P. Searles and R. Berger. Westview.

Severinghaus, C. 1955. Some observations on the breeding behavior of deer. *New York Fish and Game Journal* 2: 239–241.

Seymour, N., and R. Titman. 1980. Behavior of unpaired male black ducks *Anas rubipes* during the breeding season in a Nova Scotia Canada tidal marsh. *Canadian Journal of Zoology* 57: 2421–2428.

Shalit, R. 1993. Is rape a hate crime? *San Jose Mercury News*, June 29.

Shepard, R. 1992. The perceptual organization of colors: An adaptation to regularities of the terrestrial world. In *The Adapted Mind*, ed. J. Barkow et al. Oxford University Press.

Shields, W., and L. Shields. 1983. Forcible rape: An evolutionary perspective. *Ethology and Sociobiology* 4: 115–136.

Simpson, G. 1966. The biological nature of man. *Science* 152: 472–478.

Singh, D. 1993. Adaptive significance of female physical attractiveness: Role of waist-to-hip ratio. *Journal of Personality and Social Psychology* 59: 1192–1201.

Smith, D., and R. Prokopy. 1980. Mating behavior of *Rhagoletis pomonella* (Diptera, Tephritidae), VI. Site of early-season encounters. *Canadian Entomologist* 121: 585–590.

Smithyman, S. 1978. The Undetected Rapist. Doctoral dissertation, Claremont Graduate School. University Microfilms International.

Smuts, B. 1992. Male aggression against women: An evolutionary perspective. *Human Nature* 3: 1–44.

Smuts, B., and R. Smuts. 1993. Male aggression and sexual coercion of females in nonhuman primates and other mammals: Evidence and theoretical implications. *Advances in the Study of Behavior* 22: 1–63.

Snowden, C. 1997. The "nature" of sex differences: Myths of male and female. In *Feminism and Evolutionary Biology*, ed. P. Gowaty. Chapman and Hall.

Sober, E., and D. Wilson. 1998. *Unto Others: The Evolution and Psychology of Unselfish Behavior*. Harvard University Press.

Sociobiology Study Group. 1978. Sociobiology—Another biological determinism. In *The Sociobiology Debate*, ed. A. Caplan. Harper and Row.

Soltis, J., F. Mitsunaga, K. Shimizu, Y. Yanagihara, and M. Nazaki. 1997. Sexual selection in Japanese macaques. 1. Female mate choice or male sexual coercion. *Animal Behaviour* 54: 725–736.

Sorenson, L. 1994. Forced extra-pair copulation and mate guarding in the white-cheeked pintail: Tinting and trade-offs in an asynchronously breeding duck. *Animal Behaviour* 48: 519–533.

Sorenson, S., and J. White. 1992. Adult sexual assault: Overview of research. *Journal of Social Issues* 48: 1–8.

Sork, V. 1997. Quantitative genetics, feminism, and evolutionary theories of gender differences. In *Feminism and Evolutionary Biology*, ed. P. Gowaty. Chapman and Hall.

Southern Women's Writing Collective. 1990. Sex resistance in heterosexual arrangements. In *The Sexual Liberals and the Attack on Feminism*, ed. D. Leidholdt and J. Raymond. Pergamon.

Spalding, L. 1998. Florida's 1997 chemical castration law: A return to the Dark Ages. *Florida State University Law Review* 25: 117–139.

Spohn, C., and J. Horney. 1992. *Rape Law Reform: A Grassroots Revolution and Its Impact*. Plenum.

Steadman, L., and C. Palmer. 1995. Religion as an identifiable traditional behavior subject to natural selection. *Journal of Social and Evolutionary Systems* 18: 149–164.

Steadman, L., C. Palmer, and C. Tilley. 1996. The universality of ancestor worship. *Ethnology* 35: 63–76.

Sterling, A. 1995. Undressing the victim: The intersection of evidentiary and semiotic meanings of women's clothing in rape trials. *Yale Journal of Law and Feminism* 7: 87–132.

Stock, W. 1991. Feminist explanations: Male power, hostility and sexual coercion. In *Sexual Coercion*, ed. E. Grauerholz and M. Koralewski. Lexington.

Sturup, G. 1960. Sex offenses: The Scandinavian experience. Law and Contemporary Problems 25: 361–375.

Sturup, G. 1968. The treatment of sexual offenders in Hestedvester, Denmark. *Acta Psychiatric Scandinavia* Supplement 204.

Sulloway, F. 1996. *Born to Rebel: Family Conflict and Radical Genius*. Pantheon.

Sunday, S. 1985. Introduction. In *Violence against Women*, ed. S. Sunday and E. Tobach. Gordian Press.

Sunday, S., and E. Tobach, eds. 1985. *Violence against Women: A Critique of the Sociobiology of Rape*. Gordian Press.

Surbey, M. 1990. Family composition, stress and human menarche. In *The Socioendocrinology of Primate Reproduction*, ed. T. Ziegler and F. Bercovitch. Wiley-Liss.

Sussman, L., and S. Bordwell. 1981. *The Rapist File*. Chelsea House.

Svalastoga, K. 1962. Rape and social structure. *Pacific Sociological Review* 5: 48–53.

Swisher, K., and C. Wekesser, eds. 1994. *Violence against Women*. Greenhaven.

Symons, D. 1978. The question of function: Dominance and play. In *Social Play in Primates*, ed. E. Smith. Academic Press.

Symons, D. 1979. *The Evolution of Human Sexuality*. Oxford University Press.

Symons, D. 1987a. If we're all Darwinians, what's the fuss about? In *Sociobiology and Psychology*, ed. C. Crawford et al. Erlbaum.

Symons, D. 1987b. An evolutionary approach: Can Darwin's view of life shed light on human sexuality? In *Theories of Human Sexuality*, ed. J. Greer and W. O'Donahue. Plenum.

Symons, D. 1992. On the use and misuse of Darwinism in the study of human behavior. In *The Adapted Mind*, ed. J. Barkow et al. Oxford University Press.

Symons, D. 1995. Beauty is in the adaptations of the beholder: The evolutionary psychology of human female sexual attractiveness. In *Sexual Nature, Sexual Culture*, ed. P. Abramson and S. Pinkerton. University of Chicago Press.

Syzmanski, L., A. Devlin, J. Chrisler, and S. Vyse. 1993. Gender role and attitudes toward rape in male and female college students. *Sex Roles* 29: 37–57.

Tang-Martinez, Z. 1997. The curious courtship of sociobiology and feminism: A case of irreconcilable differences. In *Feminism and Evolutionary Biology*, ed. P. Gowaty. Chapman and Hall.

Thiessen, D. 1983. The unseen roots of rape: The theoretical untouchable. Paper presented at 1983 meetings of American Psychological Association); published in *Revue Européenne des Sciences Sociales* 24: 9–40 (1986).

Thiessen, D., and R. Young. 1994. Investigating sexual coercion. *Society*, March-April: 60–63.

Thompkins, T. 1995. Prosecuting rape as a war crime: Speaking the unspeakable. *Notre Dame Law Review* 70: 307–322.

Thornhill, N. 1996. Psychological adaptation to sexual coercion in victims and offenders. In *Sex, Power, Conflict*, ed. D. Buss and N. Malamuth. Oxford University Press.

Thornhill, N., and R. Thornhill. 1987. Evolutionary theory and rules of mating and marriage. In *Sociobiology and Psychology*, ed. C. Crawford et al. Erlbaum.

Thornhill, N., and R. Thornhill. 1990a. Evolutionary analysis of psychological pain of rape victims I: The effects of victim's age and marital status. *Ethology and Sociobiology* 11: 155–176.

Thornhill, N., and R. Thornhill. 1990b. Evolutionary analysis of psychological pain following rape II: The effects of stranger, friend and family member offenders. *Ethology and Sociobiology* 11: 177–193.

Thornhill, N., and R. Thornhill. 1990c. Evolutionary analysis of psychological pain following rape III: The effects of force and violence. *Aggressive Behavior* 16: 297–320.

Thornhill, N., and R. Thornhill. 1991. An evolutionary analysis of psychological pain following rape IV: The effect of the nature of the sexual act. *Journal of Comparative Psychology* 105: 243–252.

Thornhill, R. 1979. Male and female sexual selection and the evolution of mating systems in insects. In *Sexual Selection and Reproductive Competition in Insects*, ed. M. Blum and N. Blum. Academic Press.

Thornhill, R. 1980. Rape in *Panorpa* scorpionflies and a general rape hypothesis. *Animal Behavior* 28: 52–59.

Thornhill, R. 1981. *Panorpa* (Mecoptera: Panorpidae) scorpionflies: Systems for understanding resource-defense polygyny and alternative male reproductive efforts. *Annual Review of Ecology and Systematics* 12: 355–386.

Thornhill, R. 1983. Cryptic female choice and its implications in the scorpionfly *Harpobittacus nigriceps*. *American Naturalist* 122: 765–788.

Thornhill, R. 1984. Alternative hypotheses for traits believed to have evolved by sperm competition. In *Sperm Competition and the Evolution of Animal Mating Systems*, ed. R. Smith. Academic Press.

Thornhill, R. 1986. Relative parental contribution of the sexes to their offspring and the operation of sexual selection. In *Evolution of Animal Behavior*, ed. M. Nitecki and J. Kitchell. Oxford University Press.

Thornhill, R. 1987. The relative importance of intra- and interspecific competition in scorpionfly mating systems. *American Naturalist* 130: 711–729.

Thornhill, R. 1990. The study of adaptation. In *Interpretation and Explanation in the Study of Behavior*, volume 2, ed. M. Bekoff and D. Jamieson. Westview.

Thornhill, R. 1992a. Fluctuating asymmetry, interspecific aggression and male mating tactics in two species of Japanese scorpionflies. *Behavioral Ecology and Sociobiology* 30: 357–363.

Thornhill, R. 1992b. Fluctuating asymmetry and the mating system of the Japanese scorpionfly, *Panorpa japonica*. *Animal Behaviour* 44: 867–879.

Thornhill, R. 1997a. The concept of an evolved adaptation. In *Characterizing Human Psychological Adaptations*, ed. G. Bock and G. Cardew. Wiley.

Thornhill, R. 1997b. Rape-victim psychological pain revisited. In *Human Nature*, ed. L. Betzig. Oxford University Press.

Thornhill, R. 1999. The biology of human rape. *Jurimetrics* 39: 137–155.

Thornhill, R., and Alcock, J. 1983. *The Evolution of Insect Mating Systems*. Harvard University Press.

Thornhill, R., and B. Furlow. 1998. Stress and human reproductive behavior: Attractiveness, women's sexual development, postpartum depression, and baby's cry. *Advances in the Study of Behavior* 27: 319–369.

Thornhill, R., and S. Gangestad. 1993. Human facial beauty: Averageness, symmetry and parasite resistance. *Human Nature* 4: 237–269.

Thornhill, R., and S. Gangestad. 1996. The evolution of human sexuality. *Trends in Ecology and Evolution* 11: 98–102.

Thornhill, R., and S. Gangestad. 1999. The scent of symmetry: A human sex pheromone that signals fitness? *Evolution and Human Behavior* 20: 175–201.

Thornhill, R., and K. Grammer. 1999. The body and face of woman: One ornament that signals quality? *Evolution and Human Behavior* 20: 105–120.

Thornhill, R., and D. Gwynne. 1986. The evolution of sexual differences in insects. *American Scientist* 74: 382–389.

Thornhill, R., and A. Møller. 1997. Developmental stability, disease and medicine. *Biological Reviews* 72: 497–548.

Thornhill, R., and K. Sauer. 1991. The notal organ of the scorpionfly (*Panorpa vulgaris*): An adaptation to coerce mating duration. *Behavioral Ecology* 2: 156–164.

Thornhill, R., and N. Thornhill. 1983. Human rape: An evolutionary analysis. *Ethology and Sociobiology* 4: 137–173.

Thornhill, R., and N. Thornhill. 1989. The evolution of psychological pain. In *Sociobiology and the Social Sciences*, ed. R. Bell and N. Bell. Texas Tech University Press.

Thornhill, R., and N. Thornhill. 1991. Coercive sexuality of men: Is there psychological adaptation to rape? In *Sexual Coercion*, ed. E. Grauerholz and M. Koralewski. Lexington.

Thornhill, R., and N. Thornhill. 1992a. The evolutionary psychology of men's sexual coercion. *Behavioral and Brain Sciences* 15: 363–375.

Thornhill, R., and N. Thornhill. 1992b. The study of men's coercive sexuality: What course should it take? *Behavioral and Brain Sciences* 15: 404–421.

Thornhill, R., S. Gangestad, and R. Comer. 1995. Human female orgasm and mate fluctuating asymmetry. *Animal Behaviour* 50: 1601–1615.

Thornhill, R., N. Thornhill, and G. Dizinno. 1986. The biology of rape. In *Rape*, ed. S. Tomaseli and R. Porter. Blackwell.

Titman, R. 1983. Spacing and three bird flights of mallards *Anas platyrhynchos* breeding in pothole habitat. *Canadian Journal of Zoology* 61: 837–847.

Tobach, E., and B. Rosoff. 1985. Preface. In *Violence against Women*, ed. S. Sunday and E. Tobach. Gordian Press.

Tobach, E., and S. Sunday. 1985. Epilogue. In *Violence against Women*, ed. S. Sunday and E. Tobach. Gordian Press.

Tooby, J., and L. Cosmides. 1989. Evolutionary psychology and the generation of culture, part I: Theoretical considerations. *Ethology and Sociobiology* 10: 29–50.

Tooby, J., and L. Cosmides. 1990a. The past explains the present: Emotional adaptations and the structure of ancestral environments. *Ethology and Sociobiology* 11: 375–424.

Tooby, J., and L. Cosmides. 1990b. On the universality of human nature and the uniqueness of the individual: The role of genetics and adaptation. *Journal of Personality* 58: 1–67.

Tooby, J., and Cosmides, L. 1992. The psychological foundations of culture. In *The Adapted Mind*, ed. J. Barkow et al. Oxford University Press.

Torrey, M. 1995. Feminist legal scholarship on rape: A maturing look at one form of violence against women. *William and Mary Journal of Women and the Law* 2: 35–49.

Townsend, J. 1998. *What Women Want—What Men Want: Why the Sexes Still See Love and Commitment So Differently*. Oxford University Press.

Townsend, J., and G. Levy. 1990. Effects of potential partners' physical attractiveness and socioeconomic status on sexuality and partner selection. *Archives of Sexual Behavior* 19: 149–156.

Trivers, R. 1971. The evolution of reciprocal altruism. *Quarterly Review of Biology* 46: 35–57.

Trivers, R. 1972. Parental investment and sexual selection. In *Sexual Selection and the Descent of Man, 1881–1971*, ed. B. Campbell. Aldine.

Trivers, R. 1974. Parent-offspring conflict. *American Zoologist* 14: 249–264.

Trivers, R. 1985. *Social Evolution*. Benjamin/Cummings.

Trivers, R. 1999. As they would do to you. *Skeptic* 6: 81–83.

Tsang, D. 1995. Policing "perversions": Depo-Provera and John Money's new sexual order. In *Sex, Cells, and Same-Sex Desire*, ed. J. De Ceddo and D. Parker. Haworth.

Tsubaki, V., and T. Ono. 1986. Competition for territorial sites and alternative mating tactics in the dragonfly *Nsnnophya-pygmaea odonata Libelludidae. Behaviour* 97: 234–252.

Turke, P. 1990. Which humans behave adaptively, and why does it matter? *Ethology and Sociobiology* 11: 305–339.

Turnbull, C. 1965. *Wayward Servants: The Two Worlds of the African Pygmies.* Greenwood.

Vachss, A. 1994. Treating sexual offenders: The point. In *Crimes of Gender*, ed. G. McCuen. McCuen.

Van Den Assem, J. 1967. Territory in the three spine stickleback, *Gasterosteus aculeaturs* L: An experimental study in intraspecific competition. *Behaviour* supplement: 16.

Van Rhijn, J., and T. Groethuis. 1985. Biparental care and the basis for alternative bond-types among gulls with special reference to black-headed gulls. *Ardea* 73: 159–174.

Waage, J., and P. Gowaty. 1997. Myths of genetic determinism. In *Feminism and Evolutionary Biology*, ed. P. Gowaty. Chapman and Hall.

Walston, F., A. David, and B. Charlton. 1998. Sex differences in the content of persecutory delusions: A reflection of hostile threat in the ancestral environment. *Evolution and Human Behavior* 19: 257–260.

Ward, C. 1995. *Attitudes toward Rape: Feminist and Social Psychological Perspectives*. Sage.

Warner, C. 1980. *Rape and Sexual Assault: Management and Intervention.* Aspen.

Watson, P., and P. Andrews. An evolutionary theory of unipolar depression as an adaptation for overcoming constraints of the social niche. Unpublished manuscript.

Watson, P., G. Arnqvist, and R. Stallman. 1998. Sexual conflict and the energetic costs of mating and mate choice in water striders. *American Naturalist* 151: 46–58.

Waynforth, D. 1998. Fluctuating asymmetry and human male life history traits in rural Belize. *Proceedings of the Royal Society of London* B 265: 1497–1501.

Weiner, J. 1994. *The Beak of a Finch: A Story of Evolution in Our Time.* Knopf.

Weisfeld, G. 1994. Aggression and dominance in the social world of boys. In *Male Violence*, ed. J. Archer. Routledge.

Weisfeld, G., and R. Billings. 1988. Observations on adolescence. In *Sociobiological Perspectives on Human Development*, ed. K. MacDonald. Springer-Verlag.

Wells, K. 1977. The social behavior of anuran amphibians. *Animal Behaviour* 25: 663–693.

White, J., and R. Farmer. 1992. Research methods: How they shape views of sexual violence. *Journal of Social Issues* 48: 45–59.

White, J., and S. Sorenson. 1992. A sociocultural view of sexual assault: From discrepancy to diversity. *Journal of Social Issues* 48: 187–195.

Whitaker, C. 1987. Bureau of Justice Statistics Special Report: Elderly Victims. US Department of Justice.

Wiederman, M., and E. Allgeier. 1992. Gender differences in mate selection criteria: Sociobiological or socioeconomic explanation? *Ethology and Sociobiology* 13: 115–124.

Williams, G. 1957. Pleiotropy, natural selection, and the evolution of senescence. *Evolution* 11: 398–411.

Williams, G. 1966. *Adaptation and Natural Selection.* Princeton University Press.

Williams, G. 1985. A defense of reductionism in evolutionary biology. In *Oxford Surveys in Evolutionary Biology*, volume 2, ed. R. Dawkins and M. Ridley. Oxford University Press.

Williams, G. 1988. Huxley's evolution and ethics in sociobiological perspective. *Zygon* 23: 383–407.

Williams, G. 1992. Natural selection: Domains, levels and challenges. Oxford University Press.

Willie, R., and K. Beier. 1989. Castration in Germany. *Annals of Sex Research* 2: 103–133.

Willis, C., and L. Wrightsman. 1995. Effects of victims gaze behavior and prior relationship on rape culpability attributions. *Journal of Interpersonal Violence* 10: 367–377.

Wilson, D., and E. Sober. 1994. Reintroducing group selection to the human behavioral sciences. *Behavioral and Brain Sciences* 17: 585–654.

Wilson, E. 1975. *Sociobiology: The New Synthesis.* Harvard University Press.

Wilson, E. 1998. *Consilience: The Unity of Knowledge.* Knopf.

Wilson, M., and M. Daly. 1981. Differential maltreatment of girls and boys. *Victimology* 6: 249–261.

Wilson, M., and M. Daly. 1985. Competitiveness, risk taking, and violence: The young male syndrome. *Ethology and Sociobiology* 6: 59–73.

Wilson, M., and M. Daly. 1992. The man who mistook his wife for a chattel. In *The Adapted Mind*, ed. J. Barkow et al. Oxford University Press.

Wilson, M., and S. Mesnick. 1997. An empirical test of the bodyguard hypothesis. In *Feminism and Evolutionary Biology*, ed. P. Gowaty. Chapman and Hall.

Wilson, M., M. Daly, and J. Scheib. 1997. Femicide: An evolutionary psychological perspective. In *Feminism and Evolutionary Biology*, ed. P. Gowaty. Chapman and Hall.

Wittenberger, J. 1981. *Animal Social Behavior.* Wadsworth.

Wrangham, R., and D. Peterson. 1996. *Demonic Males: Apes and the Origins of Human Violence.* Houghton Mifflin.

Wright, R. 1990. The intelligence test. *New Republic*, January 29.

Wright, R. 1994. *The Moral Animal*. Vintage.

Wynne-Edwards, V. 1962. *Animal Dispersion in Relation to Social Behavior*. Hafner.

Young, R., and D. Thiessen. 1991. The Texas rape scale. *Ethology and Sociobiology* 13: 19–33.

Zahavi, A., and A. Zahavi. 1997. *The Handicap Principle: A Missing Piece of Darwin's Puzzle*. Oxford University Press.

Zillmann, D. 1998. *Connections between Sexuality and Aggression,* 2nd ed. Lawrence Erlbaum Associates.

Index